JOHN B. FRALEIGH, University of Rhode Island

PROBABILITY AND CALCULUS:
A BRIEF INTRODUCTION

ADDISON-WESLEY PUBLISHING COMPANY

Reading, Massachusetts • Menlo Park, California • London • Don Mills, Ontario

This book is in the
ADDISON-WESLEY SERIES IN MATHEMATICS

Consulting Editor:
LYNN H. LOOMIS

Preface

This text is suitable for use in a single semester course designed to introduce students to concepts from set theory, probability, and calculus. It is assumed that the student has had a year of high school algebra and geometry.

The calculus portion of the text constitutes a *brief* introduction to the concepts, techniques, and applications of differential and integral calculus. The treatment is largely confined to polynomial functions. This text does not give a suitable, college-level treatment of calculus for the future physical scientist, engineer, or mathematician.

The entire book could just about be covered in a semester with a really good class. However, it is expected that most classes will prefer to concentrate either on set theory and probability, or on analytic geometry and calculus. Such courses could be offered according to the following outlines.

Set Theory and Probability:

Chapter 1 (Sets and Counting)
Chapter 2 (Probability)
[If time remains, outside materials could be used, or the course could continue with Chapter 3 (Real Analytic Geometry) and Chapter 4 (Functions).]

Analytic Geometry and Calculus:

Sections 1, 2, 3, 6, and 7 of Chapter 1 (Sets and Counting)
Chapter 3 (Real Analytic Geometry)
Chapter 4 (Functions)
Chapter 5 (Differential Calculus)
Chapter 6 (Integral Calculus)

Kingston, Rhode Island   J. B. F.
June 1968

Contents

1 | Sets and Counting

A formal study of sets was first undertaken by Georg Cantor (1845–1918). The concept of a set is regarded by nonmathematicians as belonging to "the new math." This is an annoying misnomer, since Cantor published his main paper on sets in 1874. In this fast-moving age, nothing that old should be called "new." Set theory is new only to nonmathematicians. Mathematicians regard the concept of a set as the cornerstone of mathematics.

We should tell you at once that we shall take a back-door approach to set theory. Our presentation gives you no feeling for the front-door, axiomatic approach. In a front-door presentation, one assumes that there exist entities called *sets*, and one also assumes that these sets satisfy certain *axioms*. One then derives all other results of set theory from these axioms. The most popular axiomatization of set theory at the present time is probably the *Zermelo-Fraenkel set theory*, begun by E. Zermelo in 1908 and improved by A. Fraenkel from 1922 to 1925.

New mathematics is seldom first developed via an axiomatic approach. For example, set theory was not first formulated by the use of axioms, but was developed along the lines which we shall follow in this text. One uses an axiomatic approach in mathematics to put a mathematical discipline on a firm logical foundation, once the ideas have appeared. Expositions in the front-door, axiomatic style are hard for a mathematically inexperienced person to grasp, for ideas are often introduced with no motivation, and are treated very abstractly. In general, we shall use the back door in this book, or at least we shall look in at the back door to motivate what we do.

We have said that mathematicians regard the concept of a set as the cornerstone of mathematics. It is of necessity a somewhat weak and crumbly cornerstone, for the concept of a set is taken in mathematics as a primitive, undefined notion. You see, it is impossible to define every concept you use. Suppose, for example, one defines the term *set* by "A *set* is a specific collection of objects." One naturally asks what is meant by a *collection*. Perhaps then one defines: "A *collection* is an aggregate of things." What then is an *aggregate*? Now our language is finite, so after some time we will run out of new words to use and will have to repeat some words already questioned. The sequence of definitions is then circular and obviously worthless. Mathematicians realize that they must start with some undefined or primitive concept. At the moment they have agreed that *set* shall be such a primitive concept, and they try to base all their mathematics on the notion of a set as

1

much as they can. We shall not define *set*, but shall just hope that when such expressions as "the set of all real numbers" or "the set of all members of the United States Senate" are used, people's various ideas of what is meant are sufficiently similar to make communication feasible. We shall consider the terms *set* and *collection* to be synonymous.

1. SETS AND SUBSETS

1.1 How to Describe a Set

People frequently single out some specific collection of objects for consideration as a whole. This occurs not only in mathematics, but also during conversations, private meditations, and indeed in every type of mental activity. For example, in writing this text, we attempt to consider the set of all students who might use it. The importance of the concept of a set lies precisely in the recognition that, in some situation, a number of **elements** comprise a single entity, that is, they form a set. The generality of the notion of a set makes it clear that sets pertain to all aspects of our culture. As explained in the introduction, we do not attempt to define a set.

There are two ways to describe to someone a certain set which you have in mind. If the set has only a small, finite number of elements, you can make a list of these elements for him. The standard way to give such a list is to enclose the elements, separated by commas, between braces. The following example illustrates this method.

Example 1.1 The set $\{1, 2, 3, 4, 5\}$ contains five elements, the first five positive integers (whole numbers). The set $\{a, e, i, o, u\}$ is the set of all vowels in our alphabet. ‖

This listing technique may be extended to describe certain large finite sets and even some infinite sets by using an ellipsis, written "..." and read "and so forth." For example, $\{1, 2, 3, 4, \ldots, 1000\}$ is the set containing the first thousand positive integers (whole numbers), while $\{1, 2, 3, 4, \ldots\}$ is the infinite set consisting of *all* positive integers.

Another, more useful, technique for describing a set is to use some characterizing property of the elements, that is, to give some property such that the set consists precisely of the things which have that property. Perhaps you noticed that in Example 1.1 we gave such properties for the two sets described there. Thus the property "Each element is one of the first five positive integers" characterizes the elements of the set $\{1, 2, 3, 4, 5\}$, and the property "Each element is a vowel in our alphabet" characterizes the elements of $\{a, e, i, o, u\}$. If $P(x)$ is some assertion regarding x, we shall let

$$\{x \mid P(x)\}$$

be the set of all elements x such that $P(x)$ is true.

Example 1.2 If $P(x)$ is the assertion "x is a vowel in our alphabet," then

$$\{x \mid P(x)\} = \{a, e, i, o, u\}.$$

Also,

$$\{1, 2, 3\} = \{x \mid x \text{ is a positive integer} \leq 3\}$$
$$= \{x \mid 2x \text{ is an even positive integer} \leq 6\}$$
$$= \{x \mid x \text{ is a solution of } (x - 1)(x - 2)(x - 3) = 0\}.$$

This illustrates that often a set can be described by several different characterizing properties of its elements. ‖

In general, we shall let capital letters denote sets and lower-case letters denote elements of sets. There are standard notations to express whether or not an object x is an element of a set A. The notation

$$\text{"}x \in A\text{"}$$

means that x is an element of A, and

$$\text{"}x \notin A\text{"}$$

means that x is not in the set A. Thus if $A = \{1, 2, 3, 4, 5\}$, we have $2 \in A$, while $7 \notin A$.

Indiscriminate use of the set-building notation "$\{x \mid P(x)\}$" can create problems. Consider for example

$$\{x \mid x \text{ is a positive integer and } x < 0\}.$$

Now no positive integer is less than 0, so one is faced with the choice of either saying that no such set exists or assuming that there exists a set having no elements. Mathematicians have chosen the latter alternative, and postulate the existence of exactly one set $\varnothing$ having no elements. This set $\varnothing$ is the **empty set** or **null set.** Another problem caused by the unrestricted use of the notation "$\{x \mid P(x)\}$" is considered at the end of this section.

One must be very careful to distinguish between a *set* and an *element* of the set. The set $\{b\}$ has just one element, the element b. A set of exactly one element is a **unit set** or **singleton set.** The *set* $\{b\}$ is different from the *element b.* Perhaps some further examples will make this clearer.

Example 1.3 The set $\varnothing$ has no elements. The set $\{\varnothing\}$ has one element, namely $\varnothing$. This example also illustrates that one set may be an element of another set. ‖

Example 1.4 The set $\{1, \{2, 3\}, 4\}$ has three elements, namely 1, $\{2, 3\}$, and 4. Note that $2 \notin \{1, \{2, 3\}, 4\}$. ‖

Example 1.5 An element is either in a set or not in a set; there is no such thing as being in a set twice. Thus "$\{1, 2, 1\}$" is not a desirable notation for a set. On the other hand, "$\{1, 2, \{1\}\}$" is a perfectly good notation; the elements 1, 2, and $\{1\}$ are distinct entities. ‖

Definition. Two set are *equal* if and only if they have exactly the same elements.

1.2 Subsets

Frequently one wishes to concentrate on some portion of a set. This leads to the notion of a *subset of a set*.

Definition. A set A is a *subset of a set* B if every element of A is in B, that is, if $x \in A$ implies that $x \in B$ also. We shall denote that A is a subset of B by "$A \subseteq B$", while "$A \subset B$" shall mean $A \subseteq B$ but $A \neq B$.

It is easy to see that $A = B$ if and only if $A \subseteq B$ and $B \subseteq A$.

Example 1.6 We have $\{1, 2\} \subseteq \{1, 2, 3\}$ and $\{1, 2\} \subset \{1, 2, 3\}$. Also, $\{1, \{2\}\} \subset \{1, \{2\}, 3\}$, but $\{1, 2\}$ is not a subset of $\{1, \{2\}, 3\}$. ‖

According to our definition, it is clear that for every set A, we have

$$A \subseteq A.$$

Since $\varnothing$ contains no element *not* in the set A, it also follows logically from our definition that

$$\varnothing \subseteq A$$

for every set A. The student may simply take $\varnothing \subseteq A$ as a further definition if he prefers.

Definition. The subsets $\varnothing$ and A of a set A are *improper subsets of A*. All other subsets of A are *proper subsets of A*.

Example 1.7 Every set A except $\varnothing$ has exactly two improper subsets, $\varnothing$ and A, where $\varnothing \neq A$. Clearly, $\varnothing$ has only one subset, the improper subset $\varnothing$. ‖

Example 1.8 While $\varnothing \notin \varnothing$, we do have $\varnothing \subseteq \varnothing$. Also, we have both

$$\varnothing \in \{\varnothing\} \quad\quad \text{and} \quad\quad \varnothing \subset \{\varnothing\}. ‖$$

We hope that you are stimulated by the challenge of keeping track of exactly which category, subset versus element, you are talking about. Further practice along these lines is given in the exercises.

1.3 Russell's Paradox

Some very fine mathematical minds have questioned the assumption that we can conceive of an infinite set as a whole in the same sense that we can conceive of the set $\{1, 2, 3\}$. Cantor's work in 1874 requires us to do this, and this technique is widely used in mathematical research today. However, Kronecker (1823–1891) and Brouwer (1881–1966), and other mathematicians, have held that such procedures are unjustified and dangerous. The paradox which follows may help you to appreciate their concern.

Consider the following situation arising from unrestricted use of the notation "$\{x \mid P(x)\}$". This paradox is due to Bertrand Russell (1872–   ). If we allow $P(x)$ to be *any* property, then we can consider

$$E = \{x \mid x \notin \varnothing\},$$

or, if you prefer to leave the empty set out of it,

$$E = \{x \mid x \text{ is conceivable}\}.$$

This set E would be the "set of everything." We observe that if E is the "set of everything," then surely

$$E \in E.$$

Thus if the existence of E is granted, it is possible for a set to be an element of itself. Using $\{x \mid P(x)\}$ once more, we arrive at Russell's set,

$$\mathscr{R} = \{x \mid x \text{ is a set and } x \notin x\}.$$

For example, since $E \in E$, we have $E \notin \mathscr{R}$; however $\{1, 2, 3\} \in \mathscr{R}$ since $\{1, 2, 3\} \notin \{1, 2, 3\}$. Now try to decide whether or not $\mathscr{R} \in \mathscr{R}$. According to the definition of $\mathscr{R}$, we would have $\mathscr{R} \in \mathscr{R}$ if and only if $\mathscr{R} \notin \mathscr{R}$. This is **Russell's paradox.** Hopefully, such contradictions in the use of set theory are avoided today by using only set theory based on axioms (such as the Zermelo-Fraenkel set theory) which exclude "too large sets" such as our set E. Actually, Cantor's own work showed that given any set, one can find a larger set. This is surely inconsistent with a concept of a "set of everything," for nothing could be larger than such a set. We refer the reader to Eves and Newsom [3, Section 9.4] for a more detailed exposition along these lines.

EXERCISES

1.1 Describe each of the following sets by listing its elements, using the notation "$\{ \ , \ , \ldots , \ \}$".

a) $\{x \mid x \text{ is a number and } x^2 = 0\}$
b) $\{x \mid x \text{ is a letter in the word } farm\}$
c) $\{x \mid x \text{ is a letter in the word } happy\}$
d) $\{x \mid x \text{ is a whole number and } -1 < x < 1\}$
e) $\{x \mid x \text{ is a whole number and } -2 \leq x \leq 2\}$
f) $\{x \mid x \text{ is a number and } x^3 - 2x^2 - 3x = 0\}$

1.2 Describe each of the following sets in terms of a characterizing property, using the notation "$\{x \mid P(x)\}$". (Many answers are possible.)

a) $\{2, 4, 6, 8\}$ b) $\{-3, -2, -1, 0, 1, 2, 3\}$
c) $\{-1, 1\}$ d) $\{t, c, a\}$

1.3 Let $A = \{1, 2, 3, 4, 5, 6\}$ and let $B = \{1, 3, 6, 7, 8\}$. Describe each of the following sets by listing its elements, using the notation "$\{\ ,\ ,\ldots,\ \}$".

a) $\{x \mid x \in A \text{ and } x \in B\}$ b) $\{x \mid x \in A \text{ or } x \in B \text{ or both}\}$
c) $\{x \mid x \in A \text{ or } x \in B \text{ but not both}\}$ d) $\{x \mid x \in A \text{ but } x \notin B\}$
e) $\{x \mid x \in B \text{ but } x \notin A\}$

1.4 Let $A = \{a, b, ?, !\}$ and let $B = \{/, c, !, a\}$. Describe each of the following sets in terms of a characterizing property involving A and B, using the notation "$\{x \mid P(x)\}$".

a) $\{b, ?\}$ b) $\{a, !\}$ c) $\{a, b, ?, !, c, /\}$
d) $\{/, c\}$ e) $\{b, ?, c, /\}$

1.5 Let $A = \{1, 2, 3, 4, \ldots, 1000\}$, let $B = \{-5, -4, -3, \ldots, 3, 4, 5\}$, and let $C = \{5, 6, 7, \ldots, 2000\}$. Describe each of the following sets by listing its elements. An ellipsis may be used.

a) $\{x \mid x \in A \text{ and } x \in B\}$ b) $\{x \mid x \in A \text{ and } x \in C\}$
c) $\{x \mid x \in A \text{ but } x \notin C\}$ d) $\{x \mid x \in C \text{ but } x \notin A\}$
e) $\{x \mid x \in B \text{ or } x \in C \text{ or both}\}$ f) $\{x \mid x \in A \text{ or } x \in B \text{ but not both}\}$

1.6 Find the number of elements in each of the following sets.

a) $\{1, \{2, 3\}\}$ b) $\{1, \{2\}, 3\}$ c) $\{\{1, 2, 3\}\}$
d) $\{1, \{2, \{3\}\}\}$ e) $\{\{\{1\}, 2\}, 3\}\}$

1.7 a) Find all subsets of the set $\{1, 2, 3\}$. Which are the improper subsets?
b) Find all subsets of the set $\{1, \{2, 3\}\}$. Which are the proper subsets?

1.8 Mark each of the following true or false.

___ a) $2 \in \{1, \{2\}, 2\}$.   ___ b) $2 \subseteq \{1, \{2\}, 2\}$.
___ c) $\{2\} \in \{1, \{2\}, 2\}$.   ___ d) $\{2\} \subseteq \{1, \{2\}, 2\}$.
___ e) $\{\{2\}\} \in \{1, \{2\}, 2\}$.   ___ f) $\{\{2\}\} \subseteq \{1, \{2\}, 2\}$.
___ g) $\{\{\{2\}\}\} \subseteq \{1, \{2\}, 2\}$.

1.9 The mathematician insists that his sets be **well defined**. That is, if A is a set and x is any object, then definitely either $x \in A$ or $x \notin A$; there must be no ambiguity as to what is in a set. For each of the "sets" described, decide whether or not the "set" is well defined.

a) The set of some numbers
b) The set of positive whole numbers less than 5
c) The set of positive whole numbers less than -3
d) The set of some positive whole numbers less than 10

1.10 How many elements are there in the set $\varnothing$? in $\{\varnothing\}$? in $\{\varnothing, \{\varnothing\}\}$? in $\{\varnothing, \{\varnothing\}, \{\varnothing, \{\varnothing\}\}\}$? Describe how, starting with the empty set, one can form a set of 100 elements. (You are not really trying to "construct something from nothing," for $\varnothing$ is far from "nothing.")

1.11 a) Show that if A, B, and C are sets such that $A \subseteq B$ and $B \subseteq C$, then $A \subseteq C$.
b) Give an example illustrating that if A, B, and C are sets with $A \in B$ and $B \in C$, then it need not be the case that $A \in C$.

1.12 Find a set of three elements such that each of its elements is also a subset of the set.

2. SETS OF NUMBERS

As you would expect in a mathematics course, we shall have frequent occasion to refer to certain sets of numbers. We shall be concerned only with *real numbers*. Every real number, positive, negative, or zero, has a representation as an unending decimal. For example, $3 = 3.000000...$, $-\frac{2}{3} = -0.666666...$, and $\pi = 3.141592...$ are real numbers. Let us take care of some considerations of notation once and for all.

Z is the set of all integers (i.e., whole numbers: positive, negative, and zero). Thus $2 \in \mathbf{Z}$, $-5 \in \mathbf{Z}$, etc. The notation "**Z**" is derived from the German word *Zahl*, which means *number*.

$\mathbf{Z}^+$ is the set of all positive integers. (Zero is excluded.) Thus $2 \in \mathbf{Z}^+$, but $-5 \notin \mathbf{Z}^+$ and $0 \notin \mathbf{Z}^+$.

Q is the set of all rational numbers (i.e., quotients m/n of integers, where $n \neq 0$). Thus $\frac{2}{3} \in \mathbf{Q}$, $-\frac{7}{4} \in \mathbf{Q}$, and also $2 = \frac{2}{1} \in \mathbf{Q}$. Note that $\mathbf{Z} \subset \mathbf{Q}$.

$\mathbf{Q}^+$ is the set of all positive rational numbers. Thus $\frac{3}{4} \in \mathbf{Q}^+$, but $-\frac{7}{4} \notin \mathbf{Q}^+$.

R is the set of all real numbers. Thus $\pi \in \mathbf{R}$, $-\pi/2 \in \mathbf{R}$, and $\mathbf{Q} \subset \mathbf{R}$.

$\mathbf{R}^+$ is the set of all positive real numbers.

All these sets are subsets of **R**.

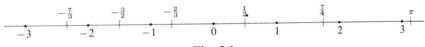

Fig. 2.1

It is often useful to visualize **R** geometrically as the set of points on the *number line*. Take a line extending infinitely in both directions. We shall make this line into an infinite ruler (see Fig. 2.1). Label any point on the line with 0 and any point to the right of 0 with 1; this fixes the scale. Each positive real number r corresponds to a point a distance r to the right of 0, while a negative number $-s$ corresponds to a point a distance s to the left of 0. For real numbers r and s, the notation "$r < s$", read "r is less than s," means that r is to the left of s on the number line. Thus $-\frac{2}{3} < \frac{1}{2}, 2 < \pi, -3 < -\frac{7}{3}$, etc. The notation "$r \leq s$" is read "$r$ is less than or equal to s."

Example 2.1 The points on the number line corresponding to

$$S = \{x \in \mathbf{R} \mid 0 \leq x \leq 2\}$$

are indicated by a heavy line in Fig. 2.2. Here we have $0 \in S$ and $2 \in S$. ‖

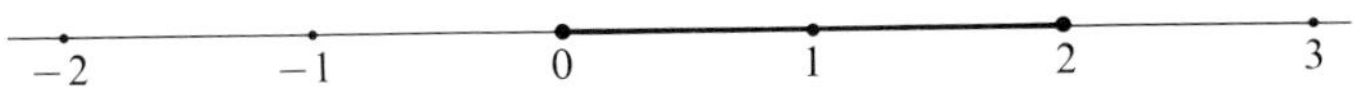

![number line from -2 to 3 with heavy line segment between 0 and 2]

Fig. 2.2

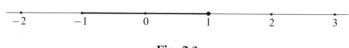

Fig. 2.3

Example 2.2 The points on the number line corresponding to

$$T = \{x \in \mathbf{R} \mid -1 < x \leq 1\}$$

are indicated by a heavy line in Fig. 2.3. Here we have $-1 \notin T$, while $1 \in T$. $\parallel$

Example 2.3 We have

$$\{x \in \mathbf{Z} \mid -\tfrac{1}{2} \leq x \leq \tfrac{3}{2}\} = \{0, 1\} \quad \text{and} \quad \{x \in \mathbf{Z}^+ \mid -\tfrac{1}{2} \leq x \leq \tfrac{3}{2}\} = \{1\}. \parallel$$

In the course of this text, we shall often have occasion to mention a set which consists of some *interval* of the number line, like the sets pictured in Figs. 2.2 and 2.3. There is a notation for such a set which is easier to write than the "$\{x \mid P(x)\}$" notation used in Examples 2.1 and 2.2.

> **Definition.** Let $a, b \in \mathbf{R}$ and let $a < b$. The ***closed interval*** $[a, b]$ ***from a to b*** is $\{x \in \mathbf{R} \mid a \leq x \leq b\}$, while $]a, b[= \{x \in \mathbf{R} \mid a < x < b\}$ is the ***open interval from a to b***. The sets $[a, b[= \{x \in \mathbf{R} \mid a \leq x < b\}$ and $]a, b] = \{x \in \mathbf{R} \mid a < x \leq b\}$ are ***half-open intervals from a to b***.

The various intervals from a to b are thus distinguished by whether neither, one, or both of the endpoints a and b belong to the interval. The brackets indicate in a natural fashion whether an endpoint is held in or cut out. The set $]a, b[$ is often denoted by "(a, b)", and similarly, many authors use "$(a, b]$" for "$]a, b]$", and "$[a, b)$" for "$[a, b[$".

Example 2.4 The interval shown in Fig. 2.2 is $[0, 2]$, and the interval shown in Fig. 2.3 is $]-1, 1]$. $\parallel$

Example 2.5 We have $1 \in]0, 2]$ and also $2 \in]0, 2]$. However, $0 \notin]0, 2]$. We also have $]0, 2] \subset [0, 2]$. $\parallel$

EXERCISES

2.1 Why do you think the notation "**Q**" is used for the set of rational numbers?

2.2 Mark each of the following true or false.

___ a) $\tfrac{2}{3} \in \mathbf{Q}^+$.   ___ b) $-2 \in \mathbf{Z}^+$.   ___ c) $-2 \in \mathbf{Z}$.

___ d) $0 \in \mathbf{Q}$.   ___ e) $0 \in \mathbf{R}^+$.   ___ f) $-\pi \in \mathbf{R}^+$.

___ g) $-\tfrac{1}{7} \in \mathbf{Z}$.   ___ h) $8 \in \mathbf{R}^+$.   ___ i) $8 \in \mathbf{Z}^+$.

___ j) $\{8\} \subset \mathbf{R}$.   ___ k) $\{\tfrac{2}{3}, 10\} \subset \mathbf{Z}$.

2.3 Mark each of the following true or false.

___ a) For all $a, b \in \mathbf{Z}^+$, we have $(a - b) \in \mathbf{Z}^+$.

___ b) For all $a, b \in \mathbf{Z}^+$, we have $(a + b) \in \mathbf{R}^+$.

___ c) For all $a \in \mathbf{Z}$, we have $2a \in \mathbf{Z}$.

___ d) For all $a \in \mathbf{Z}$, we have $a/2 \in \mathbf{Z}$.
___ e) For some $a \in \mathbf{Z}$, we have $a/2 \in \mathbf{Z}^+$.
___ f) For all $a \in \mathbf{Z}$, we have $a^2 \in \mathbf{Z}^+$.
___ g) For all $a \in \mathbf{R}$, we have $a/3 \in \mathbf{R}$.
___ h) For all nonzero $a \in \mathbf{Q}$, we have $a^2 \in \mathbf{Q}^+$.

2.4 Indicate each nonempty set which follows on a number line, as in Figs. 2.2 and 2.3.

a) $A = \{x \in \mathbf{R} \mid 2 \leq x \leq 5\}$ b) $B = \{x \in \mathbf{R} \mid 0 \leq x\}$
c) $C = \{x \in \mathbf{R} \mid x^2 = 4\}$ d) $D = \{x \in \mathbf{R} \mid x^2 < 4\}$
e) $E = \{x \in \mathbf{R} \mid x^2 \leq 4\}$ f) $F = \{x \in \mathbf{R} \mid 5 \leq x \leq -1\}$

2.5 Find all subset relations (both proper and improper) which exist between the sets A, B, C, D, E, and F of Exercise 2.4. (For example, $A \subseteq A$, $D \subset E$, etc.)

2.6 Mark each of the following true or false.

___ a) $1 \in [0, 1]$.   ___ b) $\{1\} \in [0, 2[$.
___ c) $\{1\} \subset [0, 2]$.   ___ d) $\{1\} \subset]0, 1[$.
___ e) $]0, 1[\subseteq [0, 2]$.   ___ f) $\{]0, 1[\} \subseteq [0, 2]$.
___ g) $[0, 1] \subseteq]0, 2[$.   ___ h) $[1, 2] \subseteq]0, 3[$.

2.7 Describe each nonempty set by listing its elements.

a) $\{x \in \mathbf{Z}^+ \mid x \in]0, 2]\}$ b) $\{x \in \mathbf{Z}^+ \mid x \in [-1, 1[\}$
c) $\{x \in \mathbf{Z} \mid x \in [-1, 1[\}$ d) $\{x \in \mathbf{Z} \mid x \in]0, 2]\}$

2.8 Classify each of the following intervals as open, closed, or half-open.

a) $]0, 4]$ b) $]-\frac{1}{2}, \frac{3}{2}[$ c) $[\frac{2}{3}, 5]$ d) $[3, \frac{7}{2}[$

2.9 Sketch each of the intervals given in Exercise 2.8 on a number line, as in Figs. 2.2 and 2.3.

3. SET OPERATIONS

3.1 Complements

In this section we shall discuss three natural ways to form new sets from sets already at hand. The notion of the *complement of a set* is the hardest to treat; let us take it up first. Let A be any set. The *complement of A* is roughly the set of all elements not in A. Now we saw in Section 1 that we must not allow the existence of "the set of everything," for such a set creates logical difficulties, as Russell's paradox shows. If we don't attach any meaning to "the set of everything," then surely we don't know what "the set of everything not in A" means either. *Thus when speaking of the complement of a set A, we shall always assume that A is contained within some specific, known set U which contains all elements that we wish to talk about for a given situation.* We think of U as a **universal set.** However, U is a universal set only for a particular situation, and if we deal with another situation, we may have to consider a different universal set. For example, in the first grade, the set $\{0, 1, 2, 3, \ldots\}$ of all nonnegative whole numbers is a satisfactory universal set for arithmetic lessons. However, in higher grades, an enlarged universal set containing fractions must be considered, and later the set must be further enlarged to contain all real numbers, positive, negative, and zero.

Definition. Let A be a subset of a universal set U. The ***complement*** $\bar{A}$ ***of*** A is the set of all elements not in A which are in U.

Example 3.1 Suppose the universal set for a situation is $U = \{1, 2, 3, 4, 5\}$. Then if $A = \{2, 4\}$, we have

$$\bar{A} = \{1, 3, 5\}. \parallel$$

Clearly, $\bar{U} = \varnothing$ and $\bar{\varnothing} = U$.

One can draw a suggestive picture, a *Venn diagram*, illustrating the notion of the complement of a subset A of a universal set U. Put a rectangular fence about all elements in the universal set, and within this rectangular fence, put a circular fence containing precisely the elements of the subset A, the shaded region in Fig. 3.1. The shaded region of Fig. 3.2 describes pictorially the complement of A.

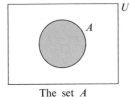

The set A

Fig. 3.1

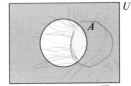

The complement $\bar{A}$

Fig. 3.2

3.2 Union and Intersection

We now introduce the operations of *union* and *intersection*. Note that the definitions of these operations given below are identical except that where the word *or* (in the inclusive sense) is used in the definition of the union, the word *and* is used in the definition of the intersection. The word *not* played a crucial role in our definition of the complement of a set. There is actually a great similarity between elementary set theory and simple formal logic. You will find this relationship explained in Kemeny, Snell, and Thompson [5, Chapters I and II].

Definition. Let A and B be sets. The ***union of*** A ***and*** B, denoted by "$A \cup B$", is the set of all elements in A or in B (or in both sets). The ***intersection of*** A ***and*** B, denoted by "$A \cap B$", is the set of all elements in A and in B.

Example 3.2 Let $A = \{1, 2, 3, 4, 7\}$ and $B = \{2, 5, 7, 10\}$. Then

$$A \cup B = \{1, 2, 3, 4, 5, 7, 10\} \qquad \text{while} \qquad A \cap B = \{2, 7\}. \parallel$$

Thus in the union $A \cup B$, *all the elements* in the sets are collected together. However, in Example 3.2 note that 2 is not listed twice in $A \cup B$. We repeat: An element is either in a set or not in a set; it is never in a set "twice," and $A \cup B$ was defined to be a set. The intersection $A \cap B$ consists of the *elements in common* to the sets.

Example 3.3 Let $A = [-1, 2[$ and let $B =]0, 3]$. Then

$$A \cup B = [-1, 3] \qquad \text{while} \qquad A \cap B =]0, 2[. \parallel$$

Consideration of the intersection of two sets provides us with another good reason for postulating the existence of the empty set $\varnothing$. Suppose that $A = \{1, 2, 3\}$ and $B = \{4, 5\}$. Then A and B have no elements in common, so $A \cap B = \varnothing$. If we did not have the empty set, we would have to say that $A \cap B$ did not exist, and every time we attempted to form an intersection of sets, we would have to worry about the existence of the intersection.

Definition. Two sets A and B are ***disjoint*** if $A \cap B = \varnothing$.

We can also describe $A \cup B$ and $A \cap B$ pictorially using Venn diagrams. Since each of the sets A and B might contain some, but perhaps not all, of the elements of the other set, we draw a Venn diagram exhibiting both A and B which covers all possible cases, as shown in Fig. 3.3. The universal set U is divided into four regions in Fig. 3.3, and the union and intersection of A and B are described pictorially by shading in Figs. 3.4 and 3.5. Any of these regions may be empty, that is, may contain no elements of U. For example, if the big outside region of Fig. 3.3 contains no elements of U, we actually have $A \cup B = U$. If A and B are disjoint, then of course there are no elements in the center region $A \cap B$. Venn diagrams will be discussed further in a later section.

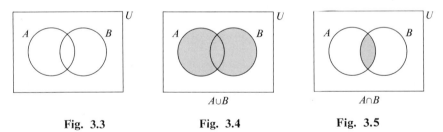

| Fig. 3.3 | Fig. 3.4 | Fig. 3.5 |

*3.3 Indexed Collections of Sets

One can form not only the union and the intersection of two sets, but also the union and intersection of any number of sets. We shall first describe a mathematical notation which is useful for handling a large number of sets.

It is not always possible to denote all sets under consideration by "A", "B", "C", etc., for after 26 sets, we run out of capital letters. The number 26 is quite small. Perhaps we actually have as many sets as there are positive integers, or even as many sets as there are real numbers. Mathematicians have invented the notation "A_i for $i \in I$" to take care of such situations. Here I, the collection of all subscripts on the A's, is some *indexing set*. For example, if we only have five sets, we can take $I = \{1, 2, 3, 4, 5\}$, and our sets are A_1, A_2, A_3, A_4, and A_5. If we have one set for each positive integer, we can consider I to be the set $\mathbf{Z}^+$, and we have sets A_1, A_2, A_3, ... where the sequence continues indefinitely. If we have one set for each real number, we can take I to be the set $\mathbf{R}$ of all real numbers, and we then have quite a lot of sets. Among them are A_{-4}, $A_{1/2}$, $A_{\sqrt{2}}$, and A_π.

With these considerations of notation out of the way, we can extend the concepts of union and intersection to cover any number of sets.

> **Definition.** Let $\{A_i \mid i \in I\}$ be a collection of sets, where I is some indexing set. Then $\bigcup_{i \in I} A_i$ is the set of all elements which are in at least one of the sets A_i, where $i \in I$, and $\bigcap_{i \in I} A_i$ is the set of all elements which are in every set A_i for $i \in I$. The set $\bigcup_{i \in I} A_i$ is the **union of the collection** $\{A_i \mid i \in I\}$ **of sets,** and $\bigcap_{i \in I} A_i$ is the **intersection of this collection.**

Clearly, this definition of union and intersection extends the one given earlier for two sets A and B. If I should be a finite set, for example, if $I = \{1, 2, \ldots, n\}$, then we may denote $\bigcup_{i \in I} A_i$ by "$A_1 \cup A_2 \cup \cdots \cup A_n$" and $\bigcap_{i \in I} A_i$ by "$A_1 \cap A_2 \cap \cdots \cap A_n$".

Don't despise discussions of notation. Modern mathematics can be very complicated, and a suggestive and easily handled notation may greatly aid in the development of new mathematics.

Example 3.4 Let $I = \mathbf{Z}^+$, and for each $i \in I$, let $A_i = [0, i]$. For example, $A_1 = [0, 1]$, $A_{10} = [0, 10]$, etc. Then

$$\bigcup_{i \in I} A_i = \{x \in \mathbf{R} \mid 0 \le x\} \qquad \text{while} \qquad \bigcap_{i \in I} A_i = A_1.$$

Note that here we have $A_1 \subset A_2 \subset A_3 \subset \cdots$ ‖

Example 3.5 Let $I = \mathbf{R}^+$, and for each $i \in I$, let $A_i = \{x \in \mathbf{Z} \mid -i < x < i\}$. For example, $A_{1/2} = \{0\}$, $A_\pi = \{-3, -2, -1, 0, 1, 2, 3\}$, etc. Then

$$\bigcup_{i \in I} A_i = \mathbf{Z} \qquad \text{while} \qquad \bigcap_{i \in I} A_i = \{0\}.$$

Note that $A_i \subseteq A_j$ for $i < j$. ‖

EXERCISES

3.1 Let the universal set U be $\{1, 2, 3, 4, 5, 6, 7, 8, 9, 10\}$. Let $A = \{2, 4, 6, 8, 10\}$, $B = \{1, 3, 5, 7, 9\}$, and $C = \{1, 2, 3, 4, 5\}$. Compute the following.

a) $\bar{A}$ b) $\bar{C}$ c) $A \cup B$ d) $A \cap B$

e) $\bar{A} \cap C$ f) $\bar{A} \cap \bar{C}$ g) $\overline{B \cup C}$ and $\bar{B} \cup \bar{C}$

h) $\overline{A \cap C}$ and $\bar{A} \cup \bar{C}$ i) $\overline{B \cup C}$ and $\bar{B} \cap \bar{C}$

3.2 Let $A = \{1, 2, 3\}$, $B = \{2, 4, 5\}$, and $C = \{3, 5, 6\}$.

a) Compute $(A \cup B) \cap C$.

b) Compute $A \cup (B \cap C)$.

c) Is the notation "$A \cup B \cap C$" ambiguous? Refer to parts (a) and (b).

3.3 Let U be the set of all students in your school. Let M be the subset of all male students, Y the subset of all students less than thirty years old, and S the subset of all single students. Describe each of the following sets symbolically.

a) The subset of all female students less than thirty years old

b) The subset of all male students at least thirty years old

c) The subset of all married students

d) The subset of all students who are either married or at least thirty years old, or both

e) The subset of all married male students less than thirty years old

f) The subset of all single female students at least thirty years old

3.4 For each of the given sets, shade in the corresponding region in a Venn diagram like that in Fig. 3.3.

a) $\overline{A \cup B}$ b) $\overline{A \cap B}$ c) $\bar{A} \cap B$

b) $A \cup \bar{B}$ e) $A \cup (\bar{A} \cap B)$

3.5 Each of the four regions in the diagram in Fig. 3.3 can be described by an intersection of two of the four sets A, $\bar{A}$, B, and $\bar{B}$. Draw a Venn diagram like that in Fig. 3.3 and label each region with the intersection that describes it.

3.6 Let A and B be subsets of a universal set U. Compute the following.

a) $A \cup \bar{A}$ b) $A \cap \bar{A}$ c) $\bar{\bar{A}}$

d) $A \cup \varnothing$ e) $A \cap \varnothing$ f) $A \cap (B \cup U)$

g) $(A \cap U) \cup A$

3.7 Let A be a set of three elements and let B be a set of five elements. Fill in the blanks.

a) $A \cup B$ has at least ___ elements. b) $A \cup B$ has at most ___ elements.

c) $A \cap B$ has at least ___ elements. d) $A \cap B$ has at most ___ elements.

3.8 Let A and B be subsets of a universal set U.

a) If $A \cup B = B$, what relationship holds between the sets A and B?

b) If $A \cap B = B$, what relationship holds between the sets A and B?

c) If $A \cup B = U$, what relationship holds between B and $\bar{A}$?

d) If $A \cap B = U$, what can be said concerning A and B?

e) If $A \cup B = \varnothing$, find $A \cup B$.

3.9 Let A and B be subsets of a universal set U.

a) What relationship must hold between A and B for the equation $A \cup X = B$ to have a solution $X \subseteq U$?

b) What relationship must hold between A and B for the equation $A \cap X = B$ to have a solution $X \subseteq U$?

3.10 Let A and B be subsets of a universal set U.

a) With reference to Exercise 3.9, show by an example that there may be more than one solution $X \subseteq U$ of $A \cup X = B$.

b) Repeat (a) for the equation $A \cap X = B$.

3.11 The *difference A − B of two sets A and B* is the set of all elements in A which are not in B. That is, $A - B = A \cap \bar{B}$.

a) Shade in the region corresponding to the set $A - B$ in a Venn diagram like that in Fig. 3.3.

b) Let $A = \{2, 4, 6, 8, 10\}$, $B = \{1, 3, 5, 7, 9\}$, and $C = \{1, 2, 3, 4, 5\}$. Compute $A - C$, $C - B$, and $A - A$.

c) For the sets in (b), compute $(A \cup B) - C$ and $A \cup (B - C)$. Is the notation "$A \cup B - C$" ambiguous?

3.12 The difference $A - B$ of sets was defined in Exercise 3.11.

a) What relationship holds between sets A and B if $A - B = A$?

b) What relationship holds between the sets A and B if $A - B = \varnothing$?

3.13 The *sum A + B of two sets A and B* is the set of all elements which are in either A or B, but which are not in both sets. That is,

$$A + B = (A \cap \bar{B}) \cup (B \cap \bar{A}).$$

(From Exercise 3.11, we see that $A + B = (A - B) \cup (B - A)$. For this reason, $A + B$ is sometimes called the *symmetric difference of A and B*.)

a) Shade in the region corresponding to the set $A + B$ in a Venn diagram like that in Fig. 3.3.

b) Let $A = \{2, 4, 6, 8, 10\}$, $B = \{1, 3, 5, 7, 9\}$, and $C = \{1, 2, 3, 4, 5\}$. Compute $A + B$, $B + C$, and $C + C$.

c) For the sets in (b), compute $(A + B) \cup C$ and $A + (B \cup C)$. Is the notation "$A + B \cup C$" ambiguous?

3.14 The sum $A + B$ of sets A and B was defined in Exercise 3.13.

a) What relationship holds between sets A and B if $A + B = A \cup B$?

b) What relationship holds between sets A and B if $A + B = \varnothing$?

c) What relationship holds between sets A and B if $A + B = U$?

*3.15 Let $A_i = \{1, 2, 3, \ldots, i\}$ for $i \in \mathbf{Z}^+$.

a) Compute $\bigcup_{i \in \mathbf{Z}^+} A_i$.   b) Compute $\bigcap_{i \in \mathbf{Z}^+} A_i$.

*3.16 Let $I = \mathbf{Z}$. For each $i \in I$, let $A_i = \{x \in \mathbf{R} \mid x < i\}$.

a) Compute $\bigcup_{i \in I} A_i$.   b) Compute $\bigcap_{i \in I} A_i$.

*3.17 Let $I = \mathbf{R}$. For each $i \in I$, let $A_i =]0, i^2]$.

a) Compute $\bigcup_{i \in I} A_i$.   b) Compute $\bigcap_{i \in I} A_i$.

*3.18 Let $I = \mathbf{R}$. For each $i \in I$, let $A_i =]i - 1, i + 2]$.

a) Compute $\bigcup_{i \in I} A_i$.   b) Compute $\bigcap_{i \in I} A_i$.

4. VENN DIAGRAMS

4.1 Number Paintings

We introduced a pictorial device, Venn diagrams, in the preceding section. In this section and the next we shall indicate how useful Venn diagrams may be, and we shall use them to demonstrate some set-theoretic identities.

We have seen that a Venn diagram which exhibits two subsets A and B of a universal set U is naturally divided into four regions. We number these

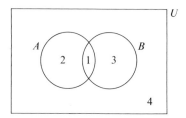

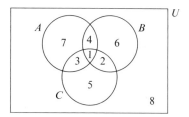

Fig. 4.1   Fig. 4.2

regions in Fig. 4.1 so that we can talk about them more easily. The choice of number for a particular region is quite immaterial. Suppose we also wish to discuss a third subset C of U. In a general situation, C may contain part of each of the four regions of Fig. 4.1, and also may exclude part of each region. Thus we should draw C so as to split each of these four regions into two parts, and we arrive at a Venn diagram with eight regions which we show with numbering in Fig. 4.2. A general Venn diagram containing a fourth subset D would contain sixteen regions, for D must be inserted in such a way that it includes and excludes portions of each of the eight regions of Fig. 4.2. This can no longer be done with another nice circle; one simply lets the boundary of D wind through any old way, so long as it splits each of the eight regions into two pieces. We leave the drawing of such a diagram to the exercises (see Exercise 4.5).

A numbering of the regions of a Venn diagram as in Fig. 4.2 may be quite useful, as the rest of this section shows.

Example 4.1 Suppose we wish to represent the set $(A \cup B) \cap \bar{C}$ by shading the appropriate regions in a three-subset Venn diagram. We can determine these regions by writing under each of the subsets A, B, and $\bar{C}$ the appropriate numbers taken from Fig. 4.2, and then proceeding as though we were taking unions, intersections, and complements of subsets of a universal set

$$\{1, 2, 3, 4, 5, 6, 7, 8\}.$$

The computation in Fig. 4.3 should now be self-explanatory. Of course, the parentheses indicate the order in which the set operations are performed. Thus the regions numbered 4, 6, and 7 in Fig. 4.2 should be shaded (Fig. 4.4). ‖

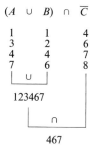

Fig. 4.3

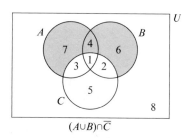

Fig. 4.4

Of course, in Example 4.1 we could shade in the correct region without the computation of Fig. 4.3 if we just thought of $(A \cup B) \cap \bar{C}$ as the set of all elements in A or B, but not in C. A more complicated expression like the one in the following example is not so easy to unravel in your head. However, the numbering technique illustrated in Example 4.1 reduces the identification of any subset of a Venn diagram to a routine computation. The author's colleagues have referred to this technique as his "number painting method."

Example 4.2 Let us shade in the set

$$[B \cup (\overline{A \cap C})] \cap [(A \cup B) \cap (\bar{B} \cup \bar{C})]$$

in a three-subset Venn diagram. We use the numbering in Fig. 4.2 again.

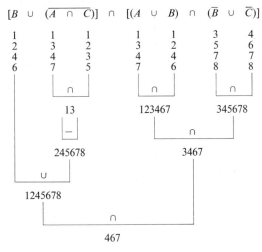

Fig. 4.5

The computation in Fig. 4.5 should now be easy for you to follow. Of course, the parentheses and brackets indicate the order in which the set operations are to be performed. Thus the regions numbered 4, 6, and 7 in Fig. 4.2 should be shaded, as shown in Fig. 4.4. ‖

4.2 Some Set-Theoretic Identities

In the preceding section you learned two ways to combine two sets A and B to form another set, namely, to form $A \cup B$ and to form $A \cap B$. You should view this as analogous to learning two ways to combine the two numbers 3 and 5 to form another number, namely, to form $3 + 5$ and to form 3×5. If you are asked to compute $2 + 3 \times 5$, you probably automatically come up with

$$2 + 3 \times 5 = 2 + 15 = 17,$$

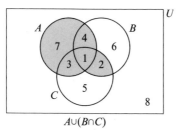 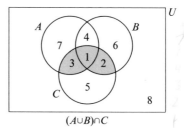

$A \cup (B \cap C)$ $(A \cup B) \cap C$

Fig. 4.6

but this is only because in algebra you were taught the *convention* that multiplication is to be performed before addition. A second grade child would be just as likely to say

$$2 + 3 \times 5 = 5 \times 5 = 25.$$

Thus the expression $2 + 3 \times 5$ is ambiguous without such a convention. In the absence of a convention, one would always have to write either "$2 + (3 \times 5)$" or "$(2 + 3) \times 5$", with parentheses. An analogous situation occurs in set theory. Subsets $A \cup (B \cap C)$ and $(A \cup B) \cap C$ of U are shaded in the Venn diagrams in Fig. 4.6. It is clear from this figure that $A \cup (B \cap C)$ does not equal $(A \cup B) \cap C$ in the event that at least one of the regions numbered 4 and 7 of the diagram is nonempty, so we must either introduce a convention for the operation to be performed first in computing an expression $A \cup B \cap C$, or always use parentheses. Here, mathematicians have *not* agreed upon any convention, and *parentheses are always used.*

We shall now develop some identities of set theory. Certain identities, such as $\overline{\overline{\varnothing}} = U$, $\overline{\overline{A}} = A$, and $A \cup \overline{A} = U$, are obvious. Others are not quite so obvious. We turn to numbers again for motivation. If a, b, and c are numbers, then ***multiplication is distributive over addition***, that is,

$$a \times (b + c) = (a \times b) + (a \times c).$$

Since $17 = 2 + (3 \times 5) \neq (2 + 3) \times (2 + 5) = 35$, it is clear that a ***distributive law for addition over multiplication***, i.e,

$$a + (b \times c) = (a + b) \times (a + c),$$

does not hold. However, in set theory, one has a superior, symmetric situation as shown by the following theorem.

Theorem 4.1 *For subsets A, B, and C of a universal set U, both distributive laws*

$$A \cap (B \cup C) = (A \cap B) \cup (A \cap C)$$

and

$$A \cup (B \cap C) = (A \cup B) \cap (A \cup C)$$

hold.

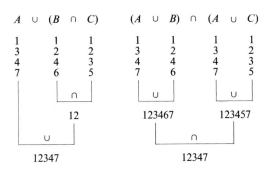

Fig. 4.7

Proof. We shall prove that

$$A \cup (B \cap C) = (A \cup B) \cap (A \cup C),$$

and leave the distributivity of intersection over union for the exercises (see Exercise 4.7). We proceed in Fig. 4.7 as in Example 4.1, taking the numbering of Fig. 4.2. In both cases, regions 1, 2, 3, 4, and 7 of the diagram in Fig. 4.2 should be shaded, so the sets are the same. ∎

Perhaps you object to the preceding proof on the grounds that one can't claim to have proved something just on the basis of a picture or diagram. Actually, the results of mathematical logic show that it is permissible to establish such set-theoretic identities via Venn diagrams.

Theorem 4.2 *For subsets A and B of a universal set U, both*

$$\overline{A \cup B} = \bar{A} \cap \bar{B} \qquad and \qquad \overline{A \cap B} = \bar{A} \cup \bar{B}$$

hold. (*These are* **DeMorgan's laws.**)

Proof. We shall demonstrate that $\overline{A \cap B} = \bar{A} \cup \bar{B}$, and leave the proof of the other equation for the exercises (see Exercise 4.8). We may use the numbering of Fig. 4.1, for only two sets are involved. The computations in Fig. 4.8 show that both $\overline{A \cap B}$ and $\bar{A} \cup \bar{B}$ correspond to the regions numbered 2, 3, and 4. Thus the sets are equal. ∎

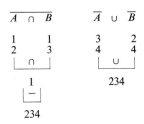

Fig. 4.8

*4.3 The Normal Form

Note that each of the numbered regions of Fig. 4.2 corresponds in a natural way to an intersection of three subsets of U. This correspondence is given by the table in Fig. 4.9. Recall that *and* was the crucial word in the definition of the intersection and *not* was the key word in the definition of the complement of a set. For example, from Fig. 4.2, we see that the region numbered 4 can be described as the region in A *and* in B *and not* in C, that is, $A \cap B \cap \bar{C}$.

1	2	3	4
$A \cap B \cap C$	$\bar{A} \cap B \cap C$	$A \cap \bar{B} \cap C$	$A \cap B \cap \bar{C}$

5	6	7	8
$\bar{A} \cap \bar{B} \cap C$	$\bar{A} \cap B \cap \bar{C}$	$A \cap \bar{B} \cap \bar{C}$	$\bar{A} \cap \bar{B} \cap \bar{C}$

Fig. 4.9

Any set obtained from A, B, and C by forming unions, intersections, and complements can be expressed as a union of sets corresponding to certain of the eight regions in Fig. 4.2. Such an expression for a set is a ***normal form for the set***. We saw in Example 4.2 that the set

$$[B \cup (\overline{A \cap C})] \cap [(A \cup B) \cap (\bar{B} \cup \bar{C})]$$

corresponds to regions 4, 6, and 7 of the diagram in Fig. 4.2. Taking the union of the sets which correspond to these regions, we arrive at the normal form for this set, namely,

$$(A \cap B \cap \bar{C}) \cup (\bar{A} \cap B \cap \bar{C}) \cup (A \cap \bar{B} \cap \bar{C}).$$

Using Theorems 4.1 and 4.2, together with $X \cup \bar{X} = U$, $X \cup X = X$, $X \cap X = X$, $X \cup \varnothing = X$, and other obvious properties of union, intersection, and complementation, we can arrive at the normal form of a set by purely algebraic means, as illustrated in the next example. We shall not emphasize this method; it is included here just for your information. In practice, we prefer to find the regions in a Venn diagram corresponding to a set, and then read off the normal form of the set from the diagram.

Example 4.3 Let us find the normal form of the set $(A \cup C) \cap (\overline{B \cup \bar{C}})$. We have

$$
\begin{aligned}
(A \cup C) \cap (\overline{B \cup \bar{C}}) &= (A \cup C) \cap (\bar{B} \cap \bar{\bar{C}}) = (A \cup C) \cap (\bar{B} \cap C) \\
&= (A \cap \bar{B} \cap C) \cup (C \cap \bar{B} \cap C) \\
&= (A \cap \bar{B} \cap C) \cup (\bar{B} \cap C \cap C) \\
&= (A \cap \bar{B} \cap C) \cup (\bar{B} \cap C).
\end{aligned}
$$

The task at this point is to introduce the missing set A into $\bar{B} \cap C$ by the device $U = A \cup \bar{A}$. Thus, continuing, we have

$$(A \cap \bar{B} \cap C) \cup (\bar{B} \cap C)$$
$$= (A \cap \bar{B} \cap C) \cup (U \cap \bar{B} \cap C)$$
$$= (A \cap \bar{B} \cap C) \cup [(A \cup \bar{A}) \cap (\bar{B} \cap C)]$$
$$= (A \cap \bar{B} \cap C) \cup [(A \cap \bar{B} \cap C) \cup (\bar{A} \cap \bar{B} \cap C)]$$
$$= (A \cap \bar{B} \cap C) \cup (A \cap \bar{B} \cap C) \cup (\bar{A} \cap \bar{B} \cap C).$$

Since $X \cup X = X$, we finally obtain as the normal form

$$(A \cup C) \cap \overline{(B \cup \bar{C})} = (A \cap \bar{B} \cap C) \cup (\bar{A} \cap \bar{B} \cap C). \;\|$$

While it is not necessary to write out as many steps in a computation as we did in Example 4.3, we still personally prefer our number painting technique of Example 4.1 to the algebraic technique of Example 4.3.

EXERCISES

4.1 How many regions are there in a general Venn diagram which exhibits

a) five subsets of a universal set?
b) six subsets of a universal set?
c) n subsets of a universal set?

4.2 Shade in each of the given sets in an appropriate Venn diagram.

a) $A \cup (\overline{\bar{A} \cap B})$ b) $(A \cap \bar{B}) \cup (\overline{B \cup A})$

4.3 Shade in each of the given sets in an appropriate Venn diagram.

a) $[(A \cap \bar{B}) \cup C] \cap \overline{(A \cup \bar{B} \cup C)}$
b) $B \cup [(\bar{A} \cap (B \cup C)) \cup (A \cap \bar{C})]$

4.4 For each of the given pairs of sets, use Venn diagrams to determine whether one is a subset of the other.

a) $A \cap B$; $[(A \cup B) \cap \bar{C}] \cup [(B \cup C) \cap \bar{A}]$
b) $[(A \cup \bar{B}) \cap \bar{C}]$; $\overline{(A \cup B)} \cap \overline{(C \cup B)}$

4.5 Draw a Venn diagram which exhibits, for a general situation, four subsets A, B, C, and D of a universal set U. Number the regions obtained.

4.6 Shade in the set $[A \cup (B \cup \bar{D})] \cap [C \cup (\bar{D} \cap A)]$ in a general four-subset Venn diagram (see Exercise 4.5).

4.7 Using Venn diagrams, prove the set-theoretic identity

$$A \cap (B \cup C) = (A \cap B) \cup (A \cap C),$$

completing the proof of Theorem 4.1.

4.8 Using Venn diagrams, prove the set-theoretic identity

$$\overline{A \cup B} = \bar{A} \cap \bar{B},$$

completing the proof of Theorem 4.2.

4.9 Prove each of the following *absorption laws.*

a) $A \cup (A \cap B) = A$ b) $A \cap (A \cup B) = A$

4.10 Use Venn diagrams to show that the notation "$A \cap B \cap C$" is not ambiguous. That is, show that $A \cap (B \cap C) = (A \cap B) \cap C$. (This is the *associative law for intersection.*)

4.11 Use Venn diagrams to show that the notation "$A \cup B \cup C$" is not ambiguous. That is, show that $A \cup (B \cup C) = (A \cup B) \cup C$. (This is the *associative law for union.*)

4.12 Use Venn diagrams to verify the set-theoretic identity

$$(\bar{A} \cap B) \cap (C \cup A) = B \cap (C \cap \bar{A}).$$

4.13 Verify the identity given in Exercise 4.12 by using Theorem 4.1 and some obvious identities such as $A \cap \bar{A} = \varnothing$.

4.14 Use Venn diagrams to verify the set-theoretic identity

$$\overline{A \cap (B \cup \bar{C})} = (\bar{B} \cap C) \cup \bar{A}.$$

4.15 Verify the identity in Exercise 4.14 by using Theorem 4.2 and some obvious identities.

4.16 Use Venn diagrams to verify the set-theoretic identity

$$\overline{(B \cap \bar{C}) \cup (\bar{A} \cap \bar{C})} = (\bar{B} \cap A) \cup C.$$

4.17 Verify the identity in Exercise 4.16 by using Theorems 4.1 and 4.2.

*4.18 Find the normal form of each of the sets in Exercise 4.2, using the Venn diagrams you found for that exercise.

*4.19 Find the normal form of each of the sets in Exercise 4.3, using the Venn diagrams you found for that exercise.

*4.20 Find the normal form of the set of Example 4.3, using the numbering technique, rather than the algebraic technique.

*4.21 Find the normal form of the set $A \cap (\bar{B} \cup C)$ by purely algebraic means, that is, without thinking in terms of Venn diagrams.

*4.22 Find the normal form of the set $A \cup (\bar{B} \cap C)$ by purely algebraic means, that is, without thinking in terms of Venn diagrams.

5. COUNTING WITH VENN DIAGRAMS

Much of the remainder of this chapter is concerned with counting the number of elements in a set. Counting problems can be exceedingly complex and hard to solve. It is not our intention to present sophisticated counting techniques or to develop your facility in solving complicated counting problems. Instead we shall present certain basic methods for use in later chapters. In this section we assume that we are dealing with sets containing only a finite number of elements, that is, with finite sets.

5.1 The Number of Elements in a Union

For a set A, we shall let $n(A)$ be the number of elements in A.

Example 5.1 If $A = \{1, 3, 5, 7\}$, then $n(A) = 4$. If $B = \{1, 3, \{5, 7\}\}$, then $n(B) = 3$. Also, $n(\varnothing) = 0$. ‖

If $n(A) = 10$ and $n(B) = 5$, then since $A \cup B$ collects the elements of both sets together, one at first might guess that

$$n(A \cup B) = n(A) + n(B) = 15.$$

Our next theorem shows that this need not be true.

Theorem 5.1 *For sets A and B, the numerical equation*

$$n(A \cup B) = n(A) + n(B) - n(A \cap B) \tag{1}$$

always holds. The equation $n(A \cup B) = n(A) + n(B)$ holds if and only if $A \cap B = \varnothing$.

Proof. We proceed via the Venn diagram shown in Fig. 5.1. The numbers are just to indicate the regions under discussion, and do *not* signify the number of elements in the regions. Let us demonstrate that

$$\underset{123}{n(A \cup B)} = \underset{12}{n(A)} + \underset{13}{n(B)} - \underset{1}{n(A \cap B)}.$$

We have written under each term of this equation the numbers of the regions which correspond to the subset which appears in that term. Thus $n(A \cup B)$ counts each element in regions 1, 2, and 3 once. Looking at the right-hand side of the equation, we see that the elements in regions 2 and 3 are each counted once, and the elements of region 1 are counted twice and then subtracted once, for a net total of one time. This shows that indeed

$$n(A \cup B) = n(A) + n(B) - n(A \cap B).$$

Clearly, the term $n(A \cap B)$ can justifiably be dropped from the equation if and only if it is zero, so $n(A \cup B) = n(A) + n(B)$ if and only if $n(A \cap B) = 0$, that is, if and only if $A \cap B = \varnothing$. ∎

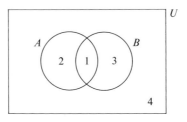

Fig. 5.1

The numerical equation

$$n(A \cup B) = n(A) + n(B) - n(A \cap B)$$

contains four numbers. In a counting problem, any three of these numbers may be given, and the fourth can then be found.

Example 5.2 Suppose that Bob had either bacon or eggs (or both) for breakfast each morning during the month of January. If he had bacon 25 mornings and eggs 18 mornings, let us find how many mornings in January he had both bacon and eggs for breakfast.

Let B be the set of those days in January on which Bob had bacon for breakfast and E the set of days on which he had eggs for breakfast. Since January has 31 days and he had either bacon or eggs (or both) each morning, we have $n(B \cup E) = 31$. Thus our problem can be phrased algebraically as

$$\underbrace{n(B \cup E)}_{31} = \underbrace{n(B)}_{25} + \underbrace{n(E)}_{18} - \underbrace{n(B \cap E)}_{?}.$$

Clearly, we must have $n(B \cap E) = 12$. Thus Bob had both bacon and eggs 12 mornings in January. ‖

The Venn diagram in Fig. 5.2 shows that the equation

$$n(A \cup B \cup C) = \underset{1347}{n(A)} + \underset{1246}{n(B)} + \underset{1235}{n(C)} - \underset{14}{n(A \cap B)}$$
$$\underset{1234567}{} $$
$$- \underset{13}{n(A \cap C)} - \underset{12}{n(B \cap C)} + \underset{1}{n(A \cap B \cap C)} \qquad (2)$$

is valid. In this equation we have written under each set the numbers corresponding to the regions in Fig. 5.2 which comprise the set. The elements in each of regions 1 through 7 are counted once on the left-hand side of the equation and a total of once on the right-hand side also. For example, on the right-hand side, the elements in region 2 are counted twice positively and once negatively, for a total of one time. Similarly, on the right-hand side, the elements of region 1 are counted three times positively, then three times negatively, and then once more positively.

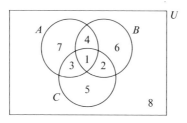

Fig. 5.2

From Eq. (1) for $n(A \cup B)$ and Eq. (2) for $n(A \cup B \cup C)$, you could probably guess correctly the analogous formula for $n(A \cup B \cup C \cup D)$. It can be shown that in general, for m subsets of a universal set, one has

$$n(A_1 \cup A_2 \cup \cdots \cup A_m) = n(A_1) + n(A_2) + \cdots + n(A_m)$$
$$- \text{ (all numbers } n(A_i \cap A_j))$$
$$+ \text{ (all numbers } n(A_i \cap A_j \cap A_k)) - \cdots$$
$$+ (-1)^{m+1} n(A_1 \cap A_2 \cap \cdots \cap A_m). \qquad (3)$$

You are asked to write out the formula for $m = 4$ in detail in Exercise 5.1.

5.2 A Digression on Special Cases versus Proofs

We wish to point out that we have *not* proved Eq. (3). We know that it holds for $m = 2$ and $m = 3$, and for $m = 1$ it reduces to $n(A) = n(A)$, which is surely true also. However, a check of a few special cases does not constitute a proof, even if the special cases seem to fall into a natural pattern. Let us give our favorite counting problem, which shows that an "obvious pattern" deduced from the first few cases of a problem need not give the correct solution for other cases.

Example 5.3 A circle with one point on its perimeter is just one region, as shown in Fig. 5.3. A circle with two points on its perimeter and the chord joining them, as in Fig. 5.4, is divided into two regions. Circles with three and four points on their perimeters, joined by all possible chords, are shown in Figs. 5.5 and 5.6. The circles are divided into four and eight regions, respectively. So far, our sequence of the number of regions goes

$$1, 2, 4, 8$$

as we increase the number of points. (The points are to be chosen so that no three of the chords meet at a single point inside the circle.) Let us now try five points, as shown in Fig. 5.7. We surely expect sixteen regions, and Fig. 5.7 verifies this. We point out that even though our sequence now is

$$1, 2, 4, 8, 16,$$

we have not *proved* that the next term, obtained by using six points on the circle and drawing all possible chords joining them, will be 32. In fact, Fig. 5.8 shows that the "obvious natural pattern" breaks down here, and the next term is 31. Isn't this a delightful example? It shows that the sequence

$$1, 2, 4, 8, 16, 31$$

is very natural, even though the people who design intelligence tests may be unaware of it. ‖

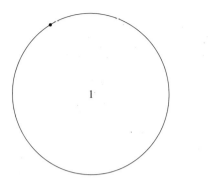

Fig. 5.3

Fig. 5.4

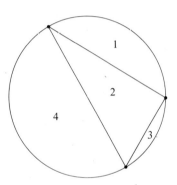

Fig. 5.5

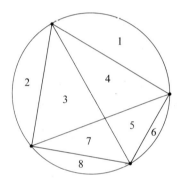

Fig. 5.6

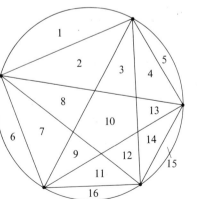

Fig. 5.7

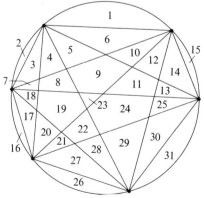

Fig. 5.8

5.3 A Popular Type of Problem

Another type of counting problem which is solvable by Venn diagrams has become popular recently in textbooks such as this one. It is best explained via an example.

Example 5.4 Consider the following problem:

In a survey of 200 *college faculty children, it is found that*

1)   68 *are well behaved,*
2) 138 *are intelligent,*
3) 160 *are talkative,*
4) 120 *are talkative and intelligent,*
5)   20 *are well behaved but not intelligent,*
6)   13 *are well behaved but not talkative,*
7)   15 *are well behaved and talkative, but not intelligent.*

How many of these 200 *children surveyed are neither well behaved, talkative, nor intelligent?*

We can consider the 200 faculty children surveyed to form our universal set U. The next thing to decide is how many different basic categories the universal set is divided into by the data. For our problem, the data can be phrased in terms of three basic subsets of U, namely,

> the subset W of well-behaved children,
> the subset I of intelligent children, and
> the subset T of talkative children.

Thus we draw a general three-subset Venn diagram, as shown in Fig. 5.9. Our problem is to find $n(\overline{W} \cap \overline{T} \cap \overline{I})$. In each of the eight regions of our Venn diagram let us place the number of children in the subset corresponding to that region. This is a bit tricky. For example, datum (1) is of no use to us at the outset, for it tells us that $n(W) = 68$, but W is divided into four regions, and we don't know how the number 68 should be split up. *We go through our data hunting for something involving all three basic categories;* in our case it is datum (7). This tells us that $n(W \cap T \cap \overline{I}) = 15$, so we place 15 in the corresponding region in Fig. 5.9. Now we hunt for a datum such that

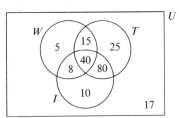

Fig. 5.9

for all but one of the regions corresponding to that datum, the numbers have been determined. For our problem, we use next datum (5), which states that $n(W \cap \bar{I}) = 20$. From the Venn diagram we see that $W \cap \bar{I}$ is composed of two regions, one of which we already know has 15 elements. Thus the other region must have 5 elements; we put a 5 in this region. We continue this process until all data have been used. In our problem, the data are used in the order (7), (5), (6), (1), (4), (2), and (3). When we finish, all our regions except the one outside the circles are numbered. Now we use the fact that $n(U) = 200$, and discover that we must have $n(\bar{W} \cap \bar{T} \cap \bar{I}) = 17$. This is the answer to our problem. ‖

The Venn diagram technique of Example 5.4 is in general much more useful than the formula technique of Example 5.2. We suggest you use Venn diagrams in the exercises.

EXERCISES

5.1 Give the explicit formula for $n(A \cup B \cup C \cup D)$ corresponding to $m = 4$ in Eq. (3).

5.2 Check your formula for $n(A \cup B \cup C \cup D)$ in Exercise 5.1, using a four-subset Venn diagram.

5.3 Determine the next term of the natural sequence 1, 2, 4, 8, 16, 31 discussed in Example 5.3.

5.4 Let $n(A) = 3$ and $n(B) = 5$. What are the possibilities for $n(A \cup B)$? How does the size of $A \cap B$ influence the size of $A \cup B$? [*Hint:* Use Eq. (1).] (Compare with Exercise 3.7.)

5.5 Bill's club meets one night a week for the 52 weeks in a year. During the year 1968, Bill never missed a meeting and played either poker or billiards (or both) each night. He played poker on 40 nights and both poker and billiards on 25 nights.

a) On how many nights at the club in 1968 did Bill play billiards?
b) On how many nights at the club in 1968 did Bill play only billiards?
c) On how many nights at the club in 1968 did Bill play only poker?

5.6 Sam belongs to the same club as Bill (see Exercise 5.5). During the year 1968, Sam never missed a night, and always played either poker or blackjack (or both) each night. Sam played just poker on 15 nights, and he played both blackjack and poker on 23 nights.

a) On how many nights at the club in 1968 did Sam play only blackjack?
b) On how many nights at the club in 1968 did Sam play blackjack?
c) On how many nights at the club in 1968 did Sam play poker?

5.7 Pete belongs to the same club as Bill and Sam (see Exercises 5.5 and 5.6). Pete always plays either blackjack or pool (or both) each night. In 1968 Pete played blackjack 22 nights and pool 30 nights. On 13 nights, Pete played both games.

a) On how many nights at the club in 1968 did Pete play only blackjack?
b) On how many nights at the club in 1968 did Pete play only pool?
c) How many meetings of his club did Pete miss in 1968?

5.8 In a survey of 200 unmarried male college students, it was found that

> 30 date intelligent girls,
> 85 date blonde girls,
> 103 date pretty girls,
> 13 date pretty, intelligent girls,
> 10 date intelligent blondes,
> 18 date pretty blondes,
> 3 date pretty, intelligent blondes.

a) Give a Venn diagram summarizing this data.
b) How many of the 200 male students surveyed date blondes who are neither pretty nor intelligent?
c) How many of the students surveyed do not date girls who are either pretty or intelligent (or both)?

5.9 A survey of 100 first grade children was taken to determine how they liked apple, grape, and orange juice. It was found that

> 63 like orange juice,
> 62 don't like apple juice,
> 18 like neither apple nor grape juice,
> 85 like at least one of the juices,
> 30 like apple and orange juice,
> 20 like all three juices,
> 28 don't like grape juice.

a) How many of the 100 children surveyed like only grape juice?
b) How many of the 100 children surveyed like both grape and apple juice?
c) How many of the 100 children surveyed like only apple juice?

5.10 Suppose the following data represents a survey of 500 students who were enrolled in a freshman college mathematics course.

> 450 passed the course,
> 10 of those who failed the course liked it,
> 25 of those who failed the course signed up
>   for another mathematics course,
> 55 of those who liked the course signed up
>   for another mathematics course,
> 60 of those who passed the course signed up
>   for another mathematics course,
> 350 of those who passed the course liked it,
> 300 of those who passed the course liked it
>   but did not sign up for another mathematics course.

a) Give a Venn diagram summarizing this data.
b) How many of the students who failed disliked the course and did not sign up for another mathematics course?
c) How many of the students liked the course?
d) How many of the students who did not like the course passed it?

5.11 The research department of a certain soap manufacturer had a field man survey 100 housewives on the use of competitors' products A, B, and C. The field man reported that of the 100 housewives

> 49 had used product A,
> 63 had used product B,
> 85 had used product C,
> 31 had used products A and B,
> 33 had used products A and C,
> 47 had used products B and C,
> 15 had used all three products.

a) Give a Venn diagram summarizing the data.
b) What should be done with this field man?

5.12 Dick belongs to the same club as Bill, Sam, and Pete (see Exercises 5.5, 5.6, and 5.7). Dick missed 3 nights in 1968. Dick always plays blackjack, cribbage, or pool (or some combination of them) each night. He never plays all three on the same night. In 1968 he played just blackjack on 8 nights, just cribbage on 5 nights, and a card game on 41 nights. He played pool on 20 nights.

a) On how many nights at the club in 1968 did Dick play both cribbage and blackjack?
b) On how many nights at the club in 1968 did Dick play something besides cribbage?
c) On how many nights at the club in 1968 did Dick play blackjack?

5.13 Show that the data $n(A) = 65$, $n(B) = 45$, $n(C) = 60$, $n(A \cap B) = 30$, $n(A \cap C) = 25$, $n(B \cap C) = 20$, and $n(A \cap B \cap C) = 25$ is inconsistent.

6. THE CARTESIAN PRODUCT

6.1 Sets of n-tuples

We now describe another important way to form a new set from given sets. Let A and B be sets. For each $a \in A$ and $b \in B$, we can form the **ordered 2-tuple** (a, b). The word *ordered* means that we shall distinguish between the 2-tuple (a, b) and the 2-tuple (b, a). For example, if both A and B are the set Z^+, then the ordered 2-tuple $(2, 3)$ is different from the ordered 2-tuple $(3, 2)$. Generalizing in the obvious way, if $A_1, A_2, \ldots, A_n$ are sets and $a_i \in A_i$, we can form the **ordered n-tuple** $(a_1, a_2, \ldots, a_n)$. Two ordered n-tuples are the same if and only if they contain the same elements in exactly the same order. For example, $(1, 5, 8) \neq (1, 8, 5)$.

> **Definition.** Let $A_1, A_2, \ldots, A_n$ be any given sets. The **Cartesian product** $A_1 \times A_2 \times \cdots \times A_n$ is the set of all ordered n-tuples $(a_1, a_2, \ldots, a_n)$ where $a_i \in A_i$.

It is also possible to generalize the concept of a Cartesian product to an infinite number of sets, but the definition is much harder to understand. We shall limit ourselves to the finite case.

Example 6.1 Let $A = \{1, 2\}$ and let $B = \{-3, 2, 4\}$. Then

$$A \times B = \{(1, -3), (1, 2), (1, 4), (2, -3), (2, 2), (2, 4)\},$$

while

$$B \times A = \{(-3, 1), (-3, 2), (2, 1), (2, 2), (4, 1), (4, 2)\}.$$

Thus, in this example, $A \times B \neq B \times A$. ‖

Often in mathematics one wishes to consider the Cartesian product of a set with itself. It is natural to let A^r be $A \times A \times \cdots \times A$ for r factors.

Example 6.2 If $A = \{1, 2\}$, then we have

$$A^2 = A \times A = \{(1, 1), (1, 2), (2, 1), (2, 2)\}. ‖$$

We have attempted to show how one tries to be careful in studying mathematics; for example, we emphasized the difference between the *element a* and the *set* $\{a\}$ in Section 1. According to our *definition*, we should distinguish between $A \times B \times C$ and $(A \times B) \times C$. Suppose, for example, that $A = \{1, 2\}$, $B = \{-3, 2, 4\}$, and $C = \{0, 5\}$. Then although we have $(1, -3, 5) \in A \times B \times C$, strictly speaking, $(1, -3, 5) \notin (A \times B) \times C$. You see, $(A \times B) \times C$ is basically a Cartesian product of *two* sets; this is the significance of the parentheses. The element of $(A \times B) \times C$ which corresponds naturally to $(1, -3, 5)$ is the ordered 2-tuple, $((1, -3), 5)$. Some practice along these lines is provided by Exercise 6.10.

6.2 Tree Diagrams and Cartesian Products

A Cartesian product of finite sets can be pictured by use of a *tree diagram*. To illustrate, let $A = \{1, 2\}$ and $B = \{-3, 2, 4\}$. If $(a, b) \in A \times B$, then the element a may be either 1 or 2, and we begin our tree by choosing a **starting point** from which we draw two **branches** labeled 1 and 2, as shown in Fig. 6.1. For each of these choices for a, the element b may be any of the three elements $-3, 2, 4$ of B. We therefore continue each of the branches in Fig. 6.1 with three branches labeled $-3, 2$, and 4, as in Fig. 6.2. Each element of $A \times B$ then corresponds to a **tree path** which commences at the starting point and continues to the right as far as it can go to a **terminal point**. For example, the top path in Fig. 6.2 corresponds to the element $(1, -3)$ of $A \times B$, the second path from the top corresponds to $(1, 2)$, etc. Clearly such a tree diagram can be made for any Cartesian product of *finite* sets.

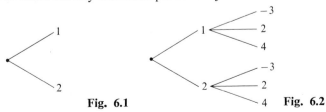

Fig. 6.1   Fig. 6.2

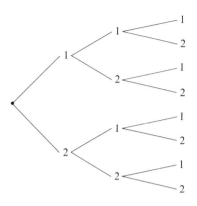

Fig. 6.3

Example 6.3 Let $A = \{1, 2\}$. The tree diagram for $A \times A \times A$ in Fig. 6.3 has eight paths. ‖

These tree diagrams suggest at once the following theorem.

Theorem 6.1 *If the set A_i has m_i elements for $i = 1, 2, \ldots, n$, then $A_1 \times A_2 \times \cdots \times A_n$ has $m_1 m_2 \cdots m_n$ elements.*

Proof. Imagine that we draw a tree diagram for $A_1 \times A_2 \times \cdots \times A_n$. There are m_1 branches emanating from the starting point. From each of these m_1 branches, there emanate m_2 branches corresponding to the m_2 elements of A_2. This gives us $m_1 m_2$ paths so far. But then each of these splits into m_3 branches corresponding to the m_3 elements of A_3, etc. The theorem is now obvious. ∎

6.3 Application to Cartesian Coordinates

There is another useful way to visualize the Cartesian product $A \times B$ of two sets A and B. We can try to arrange the elements of $A \times B$ in a rectangular pattern so that each individual element of A appears in all the first entries across just one row, and each individual element of B appears in all the second entries down just one column. To illustrate, if $A = \{1, 2\}$ and $B = \{-3, 2, 4\}$, as in Example 6.1, then $A \times B$ may be pictured as shown in Fig. 6.4. Of course, one could equally well let each individual element of A appear in all the first entries of one column, and each individual element of B appear in all the second entries of one row, as in Fig. 6.5.

	(1, 4)	(2, 4)
(1, −3) (1, 2) (1, 4)	(1, 2)	(2, 2)
(2, −3) (2, 2) (2, 4)	(1, −3)	(2, −3)

Fig. 6.4   **Fig. 6.5**

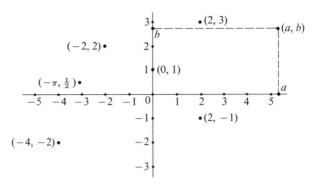

Fig. 6.6

The set **R** × **R** is the basis for *plane analytic geometry*, for if we visualize **R** × **R** as a rectangular array similar to the one in Fig. 6.5, we obtain the Euclidean plane. Recall that **R** can be viewed as a number line. We take two copies of this number line (with equal scales) and place them perpendicular to each other in a plane, so that they intersect at the point 0 on each line (see Fig. 6.6). For each element (a, b) of **R** × **R**, a point in the plane may be determined as follows. The first number a gives the position of the point with reference to the vertical number line, measured in units to the left if $a < 0$ and units to the right if $a > 0$. Similarly, the second number b gives the position of the point with reference to the horizontal number line, measured in units above if $b > 0$ and below if $b < 0$ (see Fig. 6.6). In this way, we can regard the whole plane as forming an infinite rectangular array similar to our arrangement in Fig. 6.5. This realization of the Euclidean plane as **R** × **R** is the basis for *analytic geometry* in the Euclidean plane. Each point in the plane has been provided with ***coordinates*** a and b. This device was used for the study of geometry by René Descartes (1596–1650). The term *Cartesian* is used in his honor. We shall return to analytic geometry in a later chapter.

You should think of the Euclidean plane as a picture of **R** × **R**, just as a Venn diagram is a picture of a universal set U. Any subset of **R** × **R** corresponds to a certain subset of the points in the plane. The following examples illustrate this.

Example 6.4 The subset of the Euclidean plane which corresponds to the subset

$$\{(x, y) \in \mathbf{R} \times \mathbf{R} \mid x \leq 1\}$$

of **R** × **R** is indicated in Fig. 6.7. ‖

Example 6.5 The subset of the Euclidean plane which corresponds to the subset

$$\{(x, y) \in \mathbf{R} \times \mathbf{R} \mid x \in [-2, 1] \text{ and } y \in [1, 2]\}$$

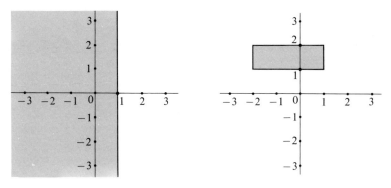

Fig. 6.7 Fig. 6.8

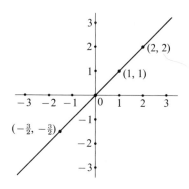

Fig. 6.9

of $\mathbf{R} \times \mathbf{R}$ is indicated in Fig. 6.8. This subset of $\mathbf{R} \times \mathbf{R}$ is easily seen to be equal to $[-2, 1] \times [1, 2]$. ‖

Example 6.6 The subset of the Euclidean plane which corresponds to the subset

$$\{(x, y) \in \mathbf{R} \times \mathbf{R} \mid x = y\}$$

of $\mathbf{R} \times \mathbf{R}$ is indicated in Fig. 6.9. ‖

EXERCISES

6.1 Let $A = \{1, 2, 3\}$ and let $B = \{1, 2\}$.

a) Draw a tree diagram for $A \times B$, as in Fig. 6.2.
b) Draw a tree diagram for $B \times A$, as in Fig. 6.2.

6.2 Mark each of the following true or false.

___ a) $(2, 3) \in \mathbf{Z} \times \mathbf{Z}$.   ___ b) $\{2, 3\} \subseteq \mathbf{R} \times \mathbf{R}$.
___ c) $\{(2, -1)\} \subseteq \mathbf{Z} \times \mathbf{R}$.   ___ d) $(-1, (3, 5)) \in \mathbf{Z} \times \mathbf{Z} \times \mathbf{Z}$.
___ e) $(-1, (3, 5)) \in \mathbf{R} \times (\mathbf{Z}^+ \times \mathbf{Z})$.   ___ f) $\mathbf{Z} \subseteq \mathbf{Z} \times \mathbf{Z}$.

___ g) $\mathbf{Z} \times \{1\} \subseteq \mathbf{Z} \times \mathbf{Z}^+$.   ___ h) $\mathbf{Z} \times \mathbf{Z} \subseteq \mathbf{Z} \times \mathbf{Z} \times \mathbf{Z}$.
___ i) $\mathbf{Z} \times \mathbf{Z} \subseteq (\mathbf{Z} \times \mathbf{Z}) \times \mathbf{Z}$.   ___ j) $[0, 3] \times \{2, 4\} \subseteq \mathbf{R} \times \mathbf{R}$.

6.3 Let A and B be finite sets. How does the tree diagram for $A \times B$ compare with the tree diagram for $B \times A$?

6.4 Let $A = \{1, 3, 5\}$ and let $B = \{2, 4, 6, 8\}$. List the elements of $A \times B$ in a rectangular array as in

a) Fig. 6.4,   b) Fig. 6.5.

6.5 Let $A = \{1, 3, 5\}$, $B = \{2, 4, 6, 8\}$, and $C = \{-1, 6\}$. Find the number of elements in each of the following sets.

a) $A \times B$ b) $B \times A$ c) $A \times B \times C$
d) B^3 e) $(A \times C)^2$ f) $[A \times (B \times A) \times A]^3$

6.6 Fill in the blanks.

a) A 3-tuple (a, b, c) is an _____ triple of elements.
b) For any set A, we have $A \times \varnothing =$ _____.
c) In analytic geometry, each point of the Euclidean plane corresponds to an element of _____.
d) The point $(2, -3)$ is located _____ units to the _____ of the vertical number line in Fig. 6.6.
e) The point $(2, -3)$ is located _____ units _____ the horizontal number line in Fig. 6.6.
f) The term *Cartesian* is used in honor of _____.
g) We have $(a, b) = (b, a)$ if and only if _____.
h) The notation "A^3" denotes the set _____.

6.7 Indicate the subset of the Euclidean plane corresponding to each of the following subsets of $\mathbf{R} \times \mathbf{R}$, as we did in Figs. 6.7, 6.8, and 6.9.

a) $\{(x, y) \in \mathbf{R} \times \mathbf{R} \mid x = 1\}$
b) $\{(x, y) \in \mathbf{R} \times \mathbf{R} \mid -1 \le x \le 2\}$
c) $\{(x, y) \in \mathbf{R} \times \mathbf{R} \mid x = -1 \text{ and } -2 \le y \le 3\}$
d) $\{(x, y) \in \mathbf{R} \times \mathbf{R} \mid x = y \text{ and } -1 \le x \le 1\}$

6.8 Proceed as in Exercise 6.7.

a) $\{(x, y) \in \mathbf{R} \times \mathbf{R} \mid x \le y\}$ b) $\{(x, y) \in \mathbf{R} \times \mathbf{R} \mid x = -y\}$
c) $\{(x, y) \in \mathbf{R} \times \mathbf{R} \mid y = 2x\}$ d) $\{(x, y) \in \mathbf{R} \times \mathbf{R} \mid 2x \le y\}$

6.9 Proceed as in Exercise 6.7.

a) $[0, 2] \times [-1, 0]$ b) $[0, 2] \times \{1\}$
c) $\{-1, 0, 1\} \times [0, 2]$ d) $\{-1, 0, 1\} \times \{2, 3\}$

6.10 Let $A = \{1, 3, 5\}$ and $B = \{2, 4, 6\}$. For each element below, give a Cartesian product involving only the sets A and B which contains the element.

a) $(2, 5)$ b) $(1, 6, 3)$ c) $(1, (6, 3))$
d) $(2, (3, 5), 6)$ e) $((2, (3, 5)), 6)$ f) $(3, (2, 5, 6))$

6.11 Try to give a geometric interpretation for $\mathbf{R} \times \mathbf{R} \times \mathbf{R}$ analogous to our geometric interpretation of $\mathbf{R} \times \mathbf{R}$ as the set of points in the Euclidean plane.

6.12 The careful mathematician usually defines the **ordered 2-tuple** (a, b) to be $\{\{a\}, \{a, b\}\}$. Why should anyone want to make a definition like this? [*Hint:* Read the introduction to this chapter again, if necessary.]

6.13 Referring to Exercise 6.12, express each of the following sets in the "(x, y)" notation for an ordered 2-tuple.

a) $\{\{1\}, \{1, -3\}\}$

b) $\{\{2, 4\}, \{4\}\}$

c) $\{\{\varnothing\}, \{1, \varnothing\}\}$

d) $\{\{1\}, \{\{\varnothing\}, 1\}\}$

6.14 Referring to Exercise 6.12, express the 2-tuple $(a, (b, c))$ as a set, without using the 2-tuple notation "(b, c)".

6.15 Referring to Exercise 6.12, give a set which would be suitable to define as the ordered 3-tuple (a, b, c).

6.16 Show that if $A \subseteq C$ and $B \subseteq D$, then $A \times B$ is a subset of $C \times D$.

6.17 Let $a, b, c \in \mathbf{R}$. The *solution set* S of the linear equation $ax + by = c$ is defined by

$$S = \{(x, y) \in \mathbf{R} \times \mathbf{R} \mid ax + by = c\}.$$

Let S_1 be the solution set of

$$2x + 3y = 12 \tag{1}$$

and let S_2 be the solution set of

$$x - y = 1. \tag{2}$$

a) Find the set of all "simultaneous solutions" of Eqs. (1) and (2).
b) Define the set of part (a) in terms of S_1 and S_2.

7. MAPS

7.1 The Intuitive Approach

The notion of a *map* (or *function*) is of great importance in all branches of mathematics. Maps are used to study properties of sets and their elements. Consider, for example, the set of all people in the world; suppose we want to study the sizes of these people. We associate with each person a number which we call his "height." The larger the number, the taller the person. Thus the property *size of people* can be studied by means of a corresponding property *size of numbers*. Assigning to each person a number corresponding to his height gives a *map* of the set of all people into the set of real numbers $\mathbf{R}$. More generally, if S and T are any sets, assigning to each element of S an element of T gives a *map* of S into T. Such a map may help you to study some property of elements of S in terms of some known property of elements of T. We shall give an intuitive, naive definition of a map here, and postpone a more precise definition until later in this section.

> **Definition.** A *map* ϕ *of a set S into a set T* is a rule which assigns to each element of S an element of T. The element of T assigned by ϕ to an element s of S is denoted by "$\phi(s)$". We say that ϕ maps S into T, denoted by "$\phi: S \to T$", and that ϕ maps s into $\phi(s)$, denoted by $s \mapsto \phi(s)$".

It is sometimes useful to visualize a map $\phi: S \to T$ as being described by arrows, where from each $s \in S$ there emanates one arrow which terminates at the element $\phi(s)$ in T. A picture of this situation is given in Fig. 7.1.

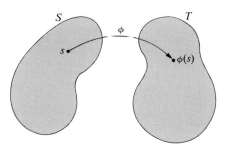

Fig. 7.1

While any symbol may be used for a map, the symbols f, g, h, ϕ, and ψ are the ones most frequently used for this purpose in mathematics.

Example 7.1 Let $f\colon \mathbf{R} \to \mathbf{R}$ be defined by $x \mapsto x^2$ for $x \notin \mathbf{R}$. Then we have $f(0) = 0^2 = 0, f(-2) = (-2)^2 = 4$, and $4 \mapsto 16$. ‖

Example 7.2 Let S be some particular set of circles. Let $\phi\colon S \to \mathbf{R}$ be defined by $\phi(s) = r$, where r is the radius of the circle s; let $\psi\colon S \to \mathbf{R}$ be defined by $\psi(s) = a$, where a is the area of the circle s; and let $f\colon S \to \mathbf{R}$ be defined by $f(s) = c$, where c is the circumference of s. Then, as is well known, $\phi(s) = r$ implies that $\psi(s) = \pi r^2$ and $f(s) = 2\pi r$. ‖

Example 7.3 Let S be the collection $\{A, B, C, \varnothing\}$ of sets where $A = \{1, 2\}$, $B = \{3, 5, 7\}$, and $C = \{1, 2, 3, 4, 6\}$. Let $n\colon S \to \mathbf{Z}$ be defined by

$$n(X) = \textit{the number of elements in } X$$

for $X \in S$. Then $n(A) = 2$, $n(B) = 3$, $n(C) = 5$, and $n(\varnothing) = 0$. We actually introduced this map n in Section 5. ‖

Example 7.4 We can describe a map ϕ of a finite set S into a set T by indicating for each element s of S the element of T which ϕ assigns to s. For example, if $S = \{a, b, c\}$ and $T = \{1, 2\}$, then

$$a \mapsto 1 \qquad b \mapsto 2 \qquad c \mapsto 1$$

describes a map ϕ of S into T. ‖

7.2 The Set-Theoretic Approach

In the introduction to this chapter we remarked that a mathematician likes to have all his ideas ultimately based on the concept of a set. Earlier in this section we defined a map to be a *rule*. What on earth is a rule? This definition is not satisfactory from a mathematical viewpoint. We would like to define a map so that *it is a set*.

Look again at Example 7.4. We defined a map $\phi\colon \{a, b, c\} \to \{1, 2\}$ by

$$a \mapsto 1 \qquad b \mapsto 2 \qquad c \mapsto 1.$$

How can we express this as a set? The concept of an ordered pair (2-tuple) is the key to this problem. We just change notation a bit and consider $a \mapsto 1$ to be the ordered pair $(a, 1)$. Similarly, $b \mapsto 2$ becomes $(b, 2)$, and $c \mapsto 1$ becomes $(c, 1)$. Our map ϕ of Example 7.4 can then be considered *to be the set* $\{(a, 1), (b, 2), (c, 1)\}$, which is a subset of the Cartesian product $\{a, b, c\} \times \{1, 2\}$. The alert student will object that we have just replaced the nebulous notion of a *rule* by the equally nebulous notion of an *ordered pair*. We point out that we showed in Exercises 6.12 through 6.15 that an ordered pair can itself be defined *to be a set*.

The preceding discussion suggests that a mathematician would consider a map $\phi\colon S \to T$ to be a subset of the Cartesian product $S \times T$. Since ϕ must assign to each element s of S a *single* element of T, for each $s \in S$, the subset of $S \times T$ must contain *exactly one* ordered pair with s as first element. We collect these ideas in a formal definition.

> **Definition.** A *map of a set S into a set T* is a subset ϕ of $S \times T$ such that each element s of S is the first member of exactly one ordered pair in ϕ.
>
> It is customary to denote $(s, t) \in \phi$ by "$\phi(s) = t$", or "$s \overset{\phi}{\mapsto} t$".

Example 7.5 Let $A = \{1, 2, 3\}$ and let $B = \{3, 5\}$. Then

$$\phi = \{(1, 5), (2, 5), (3, 3)\}$$

is a map of A into B. We have $\phi(1) = \phi(2) = 5$, while $\phi(3) = 3$. On the other hand,

$$\psi = \{(1, 5), (2, 3), (1, 3), (3, 5)\}$$

is *not* a map of A into B, for $1 \in A$ appears as the first member of *two* pairs in ψ, namely $(1, 5)$ and $(1, 3)$. Similarly,

$$f = \{(1, 5), (3, 5)\}$$

is *not* a map of A into B, for 2 does not appear as the first member of any pair in f. ‖

If ϕ is a map of A into B, that is, if $\phi\colon A \to B$, then the set A is the *domain of ϕ*. The **domain of a map** ϕ is precisely the set of all elements which appear as the first member of some ordered pair in ϕ. The subset of B which consists of all elements which appear as the second member of some ordered pair in ϕ is the **range of ϕ**.

Example 7.6 Let $\phi\colon \{a, b, c\} \to \{2, 3, 7\}$ be the map

$$\{(a, 3), (b, 7), (c, 3)\}.$$

Then $\{a, b, c\}$ is the domain of ϕ and $\{3, 7\}$ is the range of ϕ. ‖

EXERCISES

7.1 Let $\phi: \mathbf{R} \to \mathbf{R}$ be such that $x \mapsto 2x + 4$ for all $x \in \mathbf{R}$. Compute each of the following.

a) $\phi(0)$ b) $\phi(1)$ c) $\phi(-1)$
d) $\phi(2)$ e) $\phi(-2)$ f) $\phi(\frac{1}{2})$

7.2 Let $\phi: \mathbf{R}^2 \to \mathbf{R}$ be such that $(x, y) \mapsto x^2 y$ for all (x, y) in $\mathbf{R}^2$. Compute each of the following.

a) $\phi((1, 2))$ b) $\phi((0, 1))$ c) $\phi((1, 0))$ d) $\phi((-1, 1))$

7.3 Let $A = \{1, 3, 5, 7\}$ and let $B = \{2, 4\}$. Which of the following sets are maps of A into B? What is wrong with the others?

a) $\{(1, 4), (3, 4), (5, 4), (7, 4)\}$
b) $\{(1, 2), (3, 2), (1, 4), (7, 4), (5, 4)\}$
c) $\{(3, 4), (7, 2), (5, 2), (1, 2)\}$
d) $\{(3, 2), (7, 2), (5, 2)\}$

7.4 Find the domain and range of each of the given maps.

a) $\{(1, 5), (2, -7), (-3, 5)\}$
b) $\{(\varnothing, 0), (\{\varnothing\}, 1), (\{\{\varnothing\}\}, 1)\}$
c) $\{(2, 5), (3, -4), (1, 6), (-2, 5), (7, 6)\}$

7.5 Find the range of each of the indicated maps of $\mathbf{R}$ into $\mathbf{R}$. Sketch the ranges on a number line.

a) $x \mapsto x^2$ for all $x \in \mathbf{R}$ b) $x \mapsto x^3$ for all $x \in \mathbf{R}$
c) $x \mapsto x + 1$ for all $x \in \mathbf{R}$ d) $x \mapsto x^2 + 4$ for all $x \in \mathbf{R}$

e) $x \mapsto \dfrac{1}{x^2 + 1}$ for all $x \in \mathbf{R}$

7.6 If A and B are finite sets and a map $\phi: A \to B$ exists, what can you deduce about the relative sizes of the numbers $n(A)$ and $n(B)$?

7.7 If A is a set of just one element and B is a set of r elements, how many different maps are there of A into B?

7.8 Let $\phi: \{1, 3, 4, 7\} \to \{2, 3, 5\}$ be the map given by

$$1 \mapsto 3 \qquad 3 \mapsto 5 \qquad 4 \mapsto 5 \qquad 7 \mapsto 2.$$

Find the set ϕ.

7.9 If ϕ is a map such that $(2, 3) \in \phi$ and $(3, -3) \in \phi$, then compute

a) $\phi(2)$,   b) $\phi(3)$,   c) $\phi(\phi(2))$.

7.10 Consider $\phi: \mathbf{R} \to \mathbf{R}$ defined by $x \overset{\phi}{\mapsto} x^2$ and consider also $\psi: \mathbf{R} \to \mathbf{R}$ defined by $x \overset{\psi}{\mapsto} 2x - 3$. Compute the following.

a) $\phi(\psi(0))$ b) $\psi(\phi(0))$ c) $\phi(\phi(0))$
d) $\psi(\psi(0))$ e) $\phi(\psi(\phi(1)))$

7.11 Which of the following sets can be considered to be maps? What is wrong with the others? Give the domain and range of each map.

a) $\{((1, 2), 3), (1, (2, 3))\}$
b) $\{((1, 2), 3), (1, 2, 3), (1, (2, 3))\}$

c) $\{((1, 2), (1, 2)), (3, (1, 2, 4)), ((2, 4, 1), 3)\}$

d) $\{((1, 2), 3), (1, (2, 3)), (1, (3, 2))\}$

e) $\{((1, 2), 3), ((2, 1), 3), (2, (1, 3))\}$

f) $\{((1, 2), 3), (1, (2, 3, 4)), ((1, 2), (3, 4))\}$

7.12 As Exercise 7.10 suggests, if ϕ and ψ are maps such that $\phi: S \to T$ and $\psi: T \to W$, then there is a natural map of S into W, which is usually denoted by "$\psi \circ \phi$".

a) Define this natural map. That is, describe $(\psi \circ \phi)(s)$ for $s \in S$.

b) Given $S = \{1, 2\}$, $T = \{2, 3, 4\}$, $W = \{3, 5\}$, $\phi = \{(1, 3), (2, 4)\}$, and $\psi = \{(2, 5), (3, 3), (4, 3)\}$, find the set $\psi \circ \phi$.

8. TREE DIAGRAMS AND COUNTING

8.1 Sequential Events and Trees

Almost everything that you do can be broken down into a sequence of events performed in succession. For example, you don't type the five-letter word *think* all at once. The typing of this word is broken down into a sequence of actions or events; you type in succession t, h, i, n, and k. The typing of the single letter t can be broken down further into a sequence of events, namely, you position a finger over the t-key, press the key down, and let it up. Frequently, one is interested in the number of different sequences of events which can occur in some situation. For example, suppose you wish to determine the number of different five-letter words which can be formed from our alphabet, including everything from *aaaaa* to *zzzzz*, even if many of them don't mean anything in our language. You can achieve this by finding the number of different *sequences* of five alphabet keys which can be struck on a typewriter. A tree diagram is often very useful for analyzing such a situation. The technique is best introduced via examples.

Example 8.1 Suppose that Mrs. Smith wishes to arrange her three children, Mary, Bill, and Sue, in a row for a picture. In how many ways can she do this?

    We can analyze this problem as a sequence of events. There are three positions

_____ _____ _____

which Mrs. Smith has to fill with the three children, Mary, Bill, and Sue; let us suppose she fills these positions in succession from left to right. The left position could be filled by Mary, Bill, or Sue, and we begin our tree by putting down a starting point from which emanate three branches labeled with the children's initials, as shown in Fig. 8.1. We thus imagine that one child has been placed in the left position, and Fig. 8.1 shows the three possibilities.

Fig. 8.1

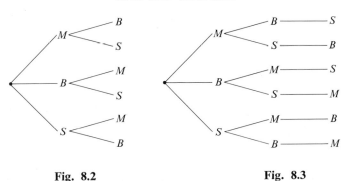

Fig. 8.2   Fig. 8.3

Now consider the middle position. If Mrs. Smith put Mary in the left position, that is, if she followed the top branch of our tree in Fig. 8.1, then either Bill or Sue could be put in the middle position. Therefore, we continue from M in Fig. 8.1 with two branches, labeled B and S, as shown in Fig. 8.2. The two lower branches of Fig. 8.1 are continued similarly, as shown in Fig. 8.2. Finally, for the right position, if Mary is at the left and Bill is in the middle, as shown on the top path of Fig. 8.2, then Sue is the only remaining child to place in the right position. Hence we continue our top path of Fig. 8.2 with only one branch, as shown in Fig. 8.3. The rest of Fig. 8.2 is similarly completed. The tree in Fig. 8.3 then has six terminal points. Recall that a path of a tree consists of one sequence of branches leading from the starting point to a terminal point. *The number of paths is always the same as the number of terminal points.* Each path of our tree represents one possibility, that is, one left-to-right order of the three children. Thus there are six possibilities in all. ‖

It is often very hard and somewhat cumbersome to describe a problem exactly, so that no misinterpretation is possible. We shall try to be reasonably explicit, but if you want to give your instructor a rough time, you can usually claim that the problem is not clearly stated and that you don't know exactly what you are asked to find. You are supposed to have an instinct for what a problem means. This is a *very* unsatisfactory situation for a mathematics course; mathematics is supposed to be the epitome of precision. However, we are dealing with applications here. We illustrate with the problem posed in Example 8.1. Mary could get into position for the picture by walking, running, hopping, skipping, jumping, rolling, turning handsprings, etc., and once there she could stand up, sit down, slouch, lie down, face front, face left, smile, frown, etc. *We were not concerned with any of these things, but only in the left-to-right order of the children as viewed by the camera.* You were supposed to sense this somehow from the statement of the problem at the start of Example 8.1.

As illustrated in Example 8.1, a tree is really a convenient picture of a *set of possibilities* for a certain situation. Thus the tree in Fig. 8.3 is really a diagram of the set

$$\{(M, B, S), (M, S, B), (B, M, S), (B, S, M), (S, M, B), (S, B, M)\}$$

of six possibilities. Clearly, if our Mrs. Smith has four children to arrange in a row, then the appropriate tree would have four branches emanating from the starting point, each of these would continue with three branches, each of these split into two branches, and finally each of those would be followed by just one branch. Thus there would be $4 \cdot 3 \cdot 2 \cdot 1 = 24$ possibilities in all. With n children, there would be

$$n(n - 1)(n - 2) \cdots (3)(2)(1) = n!$$

possibilities. This number $n(n - 1)(n - 2) \cdots (3)(2)(1)$ occurs so often that the special notation "$n!$", read "n *factorial*", is used for it. We have demonstrated the following theorem.

Theorem 8.1 *There are $n!$ ways in which n distinguishable objects can be arranged in a row.*

8.2 The Fundamental Counting Theorem

Thinking in terms of tree diagrams, we can easily prove our most fundamental counting theorem. The proof is essentially the same as our proof of Theorem 6.1 which concerned the number of elements in the Cartesian product of finite sets.

Theorem 8.2 (Fundamental counting theorem). *If in a sequence of r events, the first event has n_1 possible outcomes, and following each of these the second event has n_2 possible outcomes, etc., then the entire sequence of r events may occur in $n_1 n_2 \cdots n_r$ different ways.*

Proof. Think how you would draw a tree for such a sequence of events. Since the first event has n_1 possible outcomes, there would be n_1 first-stage branches emanating from the starting point. Since for each of these first-event outcomes, the second event has n_2 possible outcomes, there are n_2 second-stage branches emanating from the ends of each of the n_1 first-stage branches. Clearly, the number of paths thus far is $n_1 n_2$. Then from each of these $n_1 n_2$ paths, there emanate n_3 third-stage branches, etc. After r events, the total number of paths is $n_1 n_2 \cdots n_r$. ∎

Example 8.2 Let us determine the number of different five-letter words, including everything from *aaaaa* to *zzzzz*. This problem was posed at the start of the section.

We have a sequence of five slots

‾‾‾‾   ‾‾‾‾   ‾‾‾‾   ‾‾‾‾   ‾‾‾‾

to fill with letters. There are 26 choices for a letter to fill the left slot, and for each of these, there are again 26 possibilities for the next slot, etc. Thus by

Theorem 8.2, the total number of five-letter words is $26 \cdot 26 \cdot 26 \cdot 26 \cdot 26 = (26)^5 = 11{,}881{,}376$. If A is the set of lower-case letters in our alphabet, then clearly each one of these five-letter words corresponds to an element of $A^5 = A \times A \times A \times A \times A$. For example, the word *think* corresponds to (t, h, i, n, k). ‖

Example 8.3 Consider the following problem.

A coin is flipped and then a die is rolled. How many possible things can happen?

Here we have a sequence of two events. There are two possible outcomes for the first event in which a coin is flipped, namely heads, H, or tails, T. The second event in which a die is rolled has six possible outcomes, 1, 2, 3, 4, 5, or 6. Thus by Theorem 8.2, there are $2 \cdot 6 = 12$ possibilities in all. A tree showing this set of possibilities is given in Fig. 8.6. ‖

8.3 Nonsymmetric Problems

Our trees thus far have been nice and symmetric. This is not always the case. In fact, tree diagrams perhaps make their greatest contribution in situations in which they are not symmetric, as in the next example.

Example 8.4 Consider the following problem.

A bag contains two red gumdrops, one green gumdrop, and one yellow gumdrop. Gumdrops are drawn in succession without replacement until the green one is obtained, and then no more are drawn. How many possibilities are there for the sequence of colors of the gumdrops which are drawn?

The tree is shown in Fig. 8.4. Here we are interested just in *color* and not in which of the two red gumdrops may be drawn first. Thus from our starting point, we have only three first stage branches, for the three colors. If the green gumdrop is obtained on the first draw, we stop at once. In Fig. 8.4 we

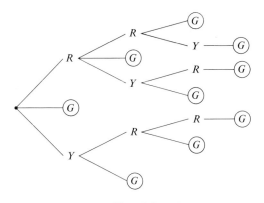

Fig. 8.4

denoted this possibility by circling the first stage G; it is a terminal point. If a red gumdrop is selected on the first draw, then there are still gumdrops of all three colors left in the bag for the second draw. If the yellow one is selected on the first draw, gumdrops of only two colors, red and green, are left in the bag. The tree is continued in this fashion. A path terminates as soon as it hits a G. Since there are nine circled terminal points, there are nine possibilities for the sequence of colors of the gumdrops which are drawn. ‖

8.4 Addition versus Multiplication

Perhaps the greatest difficulty students have with counting problems is knowing when to *add* and when to *multiply* two numbers. The answer to this question is determined by whether the English word relating the numbers is *or* (in the *exclusive* sense, i.e., not both), in which case you should add, or whether this key work is *and*, in which case you should multiply. We illustrate with two examples.

Example 8.5 Either a coin is flipped *or* a die is rolled, but not both. How many possible things can occur?

Here an appropriate tree is shown in Fig. 8.5, and this tree has $2 + 6 = 8$ paths, each representing a possibility. ‖

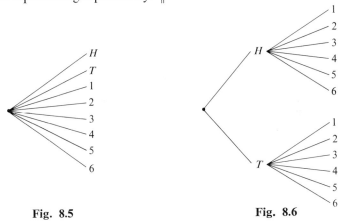

Fig. 8.5   Fig. 8.6

Example 8.6 A coin is flipped *and* then a die is rolled. How many possible things can occur?

Here the appropriate tree is shown in Fig. 8.6, and this tree has $2 \cdot 6 = 12$ paths. ‖

We state the principle just illustrated as a theorem for emphasis. We regard the proof as obvious; the multiplicative part is contained in Theorem 8.2.

Theorem 8.3 *Let one event have r possible outcomes and another event have s different possible outcomes. If one event or the other (but not both) occurs, then any of r + s things may happen. If both the one event and the other occur, then any of rs things may happen.*

EXERCISES

8.1 How many different ways can five people be arranged in a row for a picture?

8.2 How many different ways can nine people be arranged facing north in a square array of three rows and three columns?

8.3 A bag contains one red gumdrop, one yellow gumdrop, and one green gumdrop. Gumdrops are drawn in succession without replacement until the bag is empty. How many possibilities are there for the sequence of colors of the gumdrops as they are drawn? Draw a tree.

8.4 A bag contains one red gumdrop and two yellow gumdrops. Gumdrops are drawn in succession without replacement until the bag is empty. How many possibilities are there for the sequence of colors of the gumdrops as they are drawn? Draw a tree.

8.5 A bag contains two red gumdrops, one yellow gumdrop, and one green gumdrop. Gumdrops are drawn in succession without replacement until a red gumdrop is drawn, and then no more are drawn. How many possibilities are there for the sequence of colors of the gumdrops which are drawn? Draw a tree.

8.6 How many different sequences of three digits can be formed using the digits 0, 1, 2, 3, 4, 5, 6, 7, 8, 9 if repetition of digits is allowed?

8.7 Bill's birthday falls in either March or April. How many possibilities are there for the calendar date of his birthday?

8.8 Bill's birthday falls in April, and Sue's falls in May. How many possibilities are there for the combined calendar dates of their two birthdays?

8.9 A die is rolled. If an even number comes up, the die is rolled again, while if an odd number comes up, a coin is flipped. How many possible things can occur?

8.10 Either a die is rolled or a coin is flipped (but not both). If the die is rolled, then a coin is flipped. If the coin is flipped, a die is rolled if a head appears, while nothing is done if a tail appears. How many possible things can occur?

8.11 How many possibilities are there for a sequence of two flips of a coin? of three flips? of four flips? of n flips?

8.12 A coin is flipped twice, and then either the coin is flipped again or a die is rolled (but not both). How many possible things can happen?

8.13 A man has one quarter, two dimes, three nickels, and four pennies in his pocket. He selects three of these coins in succession without replacement. How many possibilities are there for the total amount of money he takes from his pocket?

8.14 A teacher decides to give two grades of A and two of B in his small seminar of four students. How many distinguishable ways can he assign these grades to his students? Draw a tree.

8.15 How many distinguishable ways can two identical white cups and two identical black cups be arranged in a row on a shelf? Draw a tree.

8.16 How many distinguishable ways can four college students be assigned to two particular vacant double rooms by the director of student housing? Draw a tree.

8.17 Bill has either orange, grape, or pineapple juice for breakfast each morning. (He never has two different kinds of juice on one morning.)

a) How many possibilities are there for the calendar menu for his breakfast juice during the month of January? (You need not multiply out your answer.)

b) How many possibilities are there for the calendar menu for his breakfast juice during the month of January if he never has the same kind of juice two days in succession?

c) How many possibilities are there for the calendar menu for his breakfast juice during the month of January if he never repeats any kind of juice until he has had each of the other two?

8.18 A mother of eight children wishes to have one help with the dishes, one go to the store, and one help with housework. How many ways can she assign children to these tasks if

a) no child is assigned two tasks?

b) any child may be assigned one, two, or all three of the tasks?

c) no child is assigned two tasks, and Billy is too small to go to the store? (Billy can do the other two jobs).

d) no child is assigned two tasks, Billy is too small to go to the store, and Sue is too clumsy to do the dishes? (Billy and Sue can each do the other two tasks.)

8.19 In how many distinguishable ways can a small club of twelve people choose three distinct members to serve as president, secretary, and treasurer?

8.20 A small club is composed of six couples. How many distinguishable ways can three distinct members be chosen to serve as president, secretary, and treasurer if it is desired to have more men than women serving in these positions?

9. TWO IMPORTANT COUNTING PROBLEMS

We are concerned in this section with two counting problems.

1) *How many subsets, both proper and improper, does a finite set of n elements have in all?*

2) *If A and B are finite sets, how many different maps are there of B into A in all?*

9.1 The Number of Subsets of a Set

Turning to question (1), let us get some insight into this problem by writing down all subsets of sets of 0, 1, 2, or 3 elements. This is done in Fig. 9.1. We see that the set $\varnothing$ of zero elements has 1 subset, a set of one element has 2 subsets, a set of two elements has 4 subsets, and a set of three elements has 8 subsets. Our natural conjecture is that a set of four elements has 16 subsets,

Set	$\varnothing$	$\{a\}$	$\{a, b\}$	$\{a, b, c\}$
Subsets	$\varnothing$	$\{a\}$ $\varnothing$	$\{a, b\}$ $\{a\}, \{b\}$ $\varnothing$	$\{a, b, c\}$ $\{a, b\}, \{a, c\}, \{b, c\}$ $\{a\}, \{b\}, \{c\}$ $\varnothing$

Fig. 9.1

but Example 5.3 should make us cautious. Actually, our guess here is correct, as the following theorem shows.

Theorem 9.1 *A set of n elements has 2^n subsets in all.*

Proof. Let $A = \{a_1, a_2, \ldots, a_n\}$ be a set of n elements. Imagine that we have an empty box in which we wish to put a subset of A. We can select a subset of A for our box by taking each element of A in turn and deciding whether or not to put it in the box, that is, whether or not to include it in the subset. We describe this process for $n = 1, 2,$ and 3 in the trees shown in Figs. 9.2, 9.3, and 9.4, where an a_i at the end of a branch means that a_i is put in the box, and an $\bar{a}_i$ means that a_i is left out of the box. For example, the two paths in Fig. 9.2 correspond, for the case $n = 1$, to the two possible subsets of the set $\{a_1\}$; the top path corresponds to the subset $\{a_1\}$ and the bottom path to the subset $\varnothing$. Likewise, the four paths in Fig. 9.3 correspond, for $n = 2$, to the four subsets of the set $\{a_1, a_2\}$, and the eight paths in Fig. 9.4 correspond, for $n = 3$, to the eight subsets of $\{a_1, a_2, a_3\}$. The top path always corresponds to the entire set, and the bottom path to the empty subset. Clearly, adding another element to the set A results in a continuation of each path with two more branches. That is, if another element is added to A, each subset of A contributes two subsets to the enlarged set, depending on whether or not the new element is put in the subset. Thus the number of subsets doubles as each element is added to A. Therefore if A has n elements, there are 2^n subsets in all. ∎

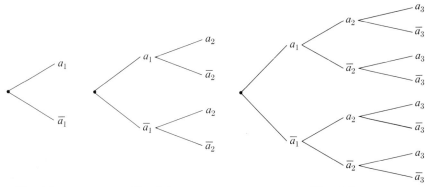

Fig. 9.2   Fig. 9.3   Fig. 9.4

9.2 The Number of Maps of B into A

Let A and B be finite nonempty sets, and let $n(A) = r$ and $n(B) = s$. We want to determine how many maps there are of B into A. We may as well let $A = \{a_1, \ldots, a_r\}$ and $B = \{b_1, \ldots, b_s\}$. A map of B into A must assign to each element of B an element of A. To see how many such maps we can define, let us select each element of B in turn, and consider the number of choices we have for an element of A to assign to that element of B. To the element $b_1 \in B$, any of the r elements of A can be assigned, *and* for each such assignment, any of the r elements of A can be assigned to $b_2 \in B$, *and* for each such assignment, any of the r elements of A can be assigned to $b_3 \in B$, etc. By Theorem 8.2, we should multiply, and we obtain

$$\underbrace{(r)(r) \cdots (r)}_{s \text{ factors}} = r^s = n(A)^{n(B)}$$

as the number of maps of B into A.

In view of this result, mathematicians let A^B be the set of all maps of B into A, where A and B may be any sets. The preceding work can then be summarized by

$$n(A^B) = n(A)^{n(B)}$$

for nonempty finite sets A and B. In our arguments we assumed that both A and B were nonempty sets. This is the most important case. The cases in which at least one of the two sets is empty are considered in the exercises (see Exercises 9.14 through 9.16). We have demonstrated the following theorem.

Theorem 9.2 *If A and B are nonempty finite sets, then $n(A^B) = n(A)^{n(B)}$.*

Example 9.1 We have

$$n(\{0, 1\}^{\{a,b\}}) = n(\{0, 1\})^{n(\{a,b\})} = 2^2 = 4.$$

Note that

$$\{0, 1\}^{\{a,b\}} = \{\{(a, 0), (b, 0)\}, \{(a, 0), (b, 1)\}, \{(a, 1), (b, 0)\}, \{(a, 1), (b, 1)\}\},$$

which is indeed a set of four elements. ‖

9.3 The Characteristic Function of a Subset

Theorem 9.1 tells us that if A is a finite set, then the number of subsets of A is $2^{n(A)}$. Theorem 9.2 tells us that this number of subsets of A is the same as the number of maps of A into a set of two elements. Let us take $\{0, 1\}$ as a set of two elements. Any map of A into $\{0, 1\}$, that is, any element of $\{0, 1\}^A$, naturally splits A into complementary subsets: the subset S of all elements of A mapped into 1 and the subset $\bar{S}$ of all elements of A mapped into 0. Conversely, starting with a subset S of a set A, we can define a map ϕ_S of A into $\{0, 1\}$ by

$$\phi_S(a) = \begin{cases} 1 & \text{if } a \in S, \\ 0 & \text{if } a \notin S. \end{cases}$$

This map ϕ_S is the ***characteristic function of the subset S of the set A***. Since each subset S of A corresponds in this natural fashion to an element of $\{0, 1\}^A$, we see again in this way that the number of subsets of A is simply $n(\{0, 1\}^A) = 2^{n(A)}$. Thus Theorem 9.1 can really be regarded as a corollary of Theorem 9.2.

Example 9.2 Let $A = \{a, b, c, d\}$ and let $S = \{a, c\}$. Then $\phi_S \in \{0, 1\}^A$ is given by

$$\phi_S = \{(a, 1), (b, 0), (c, 1), (d, 0)\}. \;\|$$

EXERCISES

9.1 Let A be a set of 7 elements.

a) How many subsets does A have?
b) How many proper subsets does A have?

9.2 Let A be a set of 2 elements.

a) How many subsets are there of $A \times A$?
b) How many subsets are there of $A \times A \times A$?

9.3 Let $A = \{1, 2, 3, 4, 5, 6\}$

a) How many subsets of A contain no odd numbers?
b) How many subsets of A contain at least one odd number?
c) How many subsets of A contain at least one even number and at least one odd number?

9.4 Let A be a set of two elements and B a set of three elements.

a) How many elements are there in A^B?
b) How many elements are there in B^A?
c) How many subsets does A^B have?
d) How many subsets does B^A have?

9.5 Let $A = \{1\}$ and let $B = \{a, b, c\}$.

a) Give the set A^B, as we did in Example 9.1.
b) Give the set B^A, as we did in Example 9.1.

9.6 Let $A = \{0, 1\}$, $B = \{1, 2, 3\}$, and $C = \{2, 3, 4, 5\}$. Find the number of elements in each of the following sets.

a) A^B b) A^A c) B^C d) $A^{B \cup C}$

9.7 Let A, B, and C be as in Exercise 9.6. Find the number of elements in each of the following sets.

a) $A^{B \cap C}$ b) $(A \times B)^B$ c) $A \times (A^B)$ d) $A^{A \times B}$

9.8 Let A, B, and C be as in Exercise 9.6. Find the number of elements in each of the following sets.

a) $(A^B)^A$ b) $(A \cap B)^C$ c) $A \cap (B^C)$ d) $B^A \cap C^A$

9.9 Let $A = \{1, 2, 3, 4, 5\}$, and consider the subset $S = \{1, 2, 5\}$ of A. Give the set ϕ_S, as we did in Example 9.2.

9.10 Let $A = \{1, 2, 3, 4, 5\}$. Give the set $\phi_{\{1,5\}}$, as we did in Example 9.2.

9.11 Let $A = \{1, 2, 3, 4, 5\}$. Give the set ϕ_A, as we did in Example 9.2.

9.12 Let $A = \{1, 2, 3, 4, 5\}$. Give the set $\phi_\varnothing$, as we did in Example 9.2.

9.13 Given that $\{(3, 0), (1, 1), (2, 0), (5, 0), (-1, 1)\}$ is the characteristic function of the subset S of a set A, find the set A and the subset S.

9.14 In arithmetic, one has $0^r = 0$ for $r \in \mathbf{Z}^+$. Show that, according to our definition of a map in Section 7.2, the formula $n(\varnothing^A) = n(\varnothing)^{n(A)}$ holds for nonempty finite sets A. [*Hint:* Remember that an element of $\varnothing^A$ is a subset of $A \times \varnothing$ which has a certain property. What is $A \times \varnothing$, and how many subsets of $A \times \varnothing$ have the property required for an element of $\varnothing^A$?]

9.15 In arithmetic, one has $r^0 = 1$ for $r \in \mathbf{Z}^+$. Show that, according to our definition of a map in Section 7.2, the formula $n(A^\varnothing) = n(A)^{n(\varnothing)}$ holds if A is a nonempty finite set. [*Hint:* Proceed as suggested in the hint for Exercise 9.14.]

9.16 In arithmetic, one usually does not define 0^0. According to our definition of a map in Section 7.2, what is $n(\varnothing^\varnothing)$? [*Hint:* Proceed as suggested in the hint of Exercise 9.14.]

†10. COUNTING SUBSETS

10.1 The General Formula

In this section we are concerned with counting how many subsets of a particular size a set has. More precisely, let S be a finite set with n elements. How many subsets of S have exactly r elements for a given r, where $0 \le r \le n$? We approach the solution to this problem via special cases.

The set $S = \{a, b, c\}$ of three elements has just three subsets of exactly two elements each: namely, $\{a, b\}$, $\{a, c\}$, and $\{b, c\}$. We can visualize this by using a tree. Suppose we wish to put exactly two elements of S into an empty box. We can accomplish this by first choosing an element of S and putting it in the box, and then choosing a second element of S and putting it in the box. The tree in Fig. 10.1 describes this process. The trouble is that each path of our tree describes not only which elements we put in the box, but also their *arrangement* in the box, that is, which element we put in first and which second. For example, in Fig. 10.1 both the top path and the third path from the top represent the same subset, $\{a, b\} = \{b, a\}$. In summary, Fig. 10.1 shows not only which two elements are in each subset, but also the $2! = 2 \cdot 1$ ways of arranging the elements in the subset. The six paths of

---

† This section is used subsequently only in the chapter on probability.

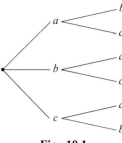

Fig. 10.1

Fig. 10.1 fall naturally into three groups of two paths each, and there are $(3 \cdot 2)/2! = 3$ subsets of $\{a, b, c\}$ containing two elements.

We illustrate with one more special case before passing to the general situation. Suppose $T = \{a_1, a_2, \ldots, a_6\}$, and suppose we wish to find the number of four-element subsets of T. Can you visualize the tree corresponding to Fig. 10.1? It is too large to draw easily. First we pick any of six elements, then any of the five remaining elements, etc., until we have selected four elements from T. The tree therefore has $6 \cdot 5 \cdot 4 \cdot 3$ paths. However, a path represents not only a subset, but also the order in which the elements were chosen in forming the subset. Since the number of different arrangements of four elements is $4!$, we see that each four-element subset of T is represented $4!$ times in the tree we are visualizing. Thus the number of subsets is

$$\frac{6 \cdot 5 \cdot 4 \cdot 3}{4 \cdot 3 \cdot 2 \cdot 1} = 15.$$

(We assume that you can cancel and do the arithmetic.)

A similar argument shows that the number of r-element subsets formed from a set of n elements, where $0 < r \leq n$, is

$$\frac{n(n - 1)(n - 2) \cdots (n - r + 1)}{r!}.$$

We denote this number by "$\binom{n}{r}$". Note that

$$\frac{n(n - 1)(n - 2) \cdots (n - r + 1)}{r!}$$

$$= \frac{n(n - 1)(n - 2) \cdots (n - r + 1)}{r!} \cdot \frac{(n - r)(n - r - 1) \cdots (3)(2)(1)}{(n - r)(n - r - 1) \cdots (3)(2)(1)}$$

$$= \frac{n!}{r!\,(n - r)!}.$$

Hence

$$\binom{n}{r} = \frac{n!}{r!\,(n-r)!}.$$

From this formula, we obtain, replacing r by $n-r$,

$$\binom{n}{n-r} = \frac{n!}{(n-r)!\,(n-(n-r))!} = \frac{n!}{(n-r)!\,r!} = \binom{n}{r}.$$

Therefore the number of r-element subsets of a set of n elements is the same as the number of $(n-r)$-element subsets. This is also clear without any computation by an argument which is illustrated in the following example.

Example 10.1 Consider the set $U = \{1, 2, 3, 4, 5\}$ of five elements. When designating a three-element subset A of U, you have the choice of designating either which elements are to be in A or which elements are *not* to be in A. That is, you can give either A or $\bar{A}$. If A contains three elements, then $\bar{A}$ contains two elements. Therefore the number of three-element subsets must be the same as the number of two-element subsets, so that $\binom{5}{3} = \binom{5}{2} = 10$. We have illustrated this in the table in Fig. 10.2. Each time you designate a subset A of a set U, you have automatically determined another subset $\bar{A}$ of U. ‖

A	$\bar{A}$	A	$\bar{A}$
$\{1, 2, 3\}$	$\{4, 5\}$	$\{1, 4, 5\}$	$\{2, 3\}$
$\{1, 2, 4\}$	$\{3, 5\}$	$\{2, 3, 4\}$	$\{1, 5\}$
$\{1, 2, 5\}$	$\{3, 4\}$	$\{2, 3, 5\}$	$\{1, 4\}$
$\{1, 3, 4\}$	$\{2, 5\}$	$\{2, 4, 5\}$	$\{1, 3\}$
$\{1, 3, 5\}$	$\{2, 4\}$	$\{3, 4, 5\}$	$\{1, 2\}$

Fig. 10.2

If we *define* $0! = 1$, then the formula

$$\binom{n}{r} = \frac{n!}{r!\,(n-r)!}$$

yields

$$\binom{n}{0} = \frac{n!}{0!\,n!} = \frac{n!}{(1)n!} = 1 \quad \text{and} \quad \binom{n}{n} = \frac{n!}{n!\,0!} = \frac{n!}{n!\,(1)} = 1.$$

Since a set S of n elements has just one subset of zero elements, namely $\varnothing$, and just one n-element subset, namely S itself, we see that the definition $0! = 1$ is indeed convenient.

We summarize our work so far in a theorem.

Theorem 10.1 *Let S be a set of n elements. Then S has $\binom{n}{r}$ subsets of exactly r elements for $0 \le r \le n$, where*

$$\binom{n}{r} = \frac{n!}{r!\,(n-r)!}.$$

Furthermore, $\binom{n}{r} = \binom{n}{n-r}$.

The student should train himself to think of $\binom{n}{r}$ as *the number of distinguishable subsets of r elements which can be formed from a set with n elements.* Do not think of $\binom{n}{r}$ as $[n!]/[r!\,(n-r)!]$; this is just a formula for computing $\binom{n}{r}$.

10.2 Some Illustrations

Quite a variety of counting problems can be solved by use of the counting results of Section 8 and Theorem 10.1. We conclude this section with some examples.

Example 10.2 Consider the following problem.

A small club has nine members. An entertainment committee of three members is to be selected. How many possibilities are there for the membership of this committee?

We are concerned here with the straightforward problem of determining the number of different subsets of three people that can be formed from a set of nine people. The answer is immediately seen to be $\binom{9}{3}$ by Theorem 10.1. Let us compute $\binom{9}{3}$. There seems to be a strong tendency for students to think that $\binom{9}{3} = \frac{9!}{3!}$. *This is wrong.* We have

$$\binom{9}{3} = \frac{9!}{3!\,(9-3)!} = \frac{9!}{3!\,6!} = \frac{9\cdot8\cdot7\cdot6\cdot5\cdot4\cdot3\cdot2\cdot1}{3\cdot2\cdot1\cdot6\cdot5\cdot4\cdot3\cdot2\cdot1} = \frac{9\cdot8\cdot7}{3\cdot2\cdot1} = 84. \parallel$$

Example 10.3 Let us modify Example 10.2 by assuming that two ladies in the club, Ann and Mary, must not both be on the entertainment committee, since they don't get along well together. Now how many possibilities are there for the membership of the committee?

We present two approaches to this question.

APPROACH 1. A committee must contain either exactly one of these two troublesome ladies *or* neither of them. Thus we are in an *or* situation, and by Theorem 8.3, we should *add* the number of committees containing just one of these ladies to the number of committees containing neither Ann nor Mary.

To form a committee containing just one of these ladies, we must select one of the two, which can be done in $\binom{2}{1}$ ways, *and* select two more people from the remaining seven, which can be done in $\binom{7}{2}$ ways. By Theorem 8.3, for this *and* situation, we should *multiply*, so the number of such committees. is $\binom{2}{1}\binom{7}{2} = 42$. (You can verify the computation.)

To form a committee containing neither Ann nor Mary, we must choose three people from the remaining seven, which can be done in $\binom{7}{3} = 35$ ways
We then *add*, arriving at $42 + 35 = 77$ for our answer.

APPROACH 2. We must not have both Ann and Mary on the committee at once. A committee containing both Ann and Mary is completely determined by the other member, and there are $\binom{7}{1} = 7$ choices for this other member. Thus there are seven committees which are not allowed. Subtracting this number from the total of 84 possible committees found in Example 10.2, we see that there are $84 - 7 = 77$ acceptable committees. ‖

The preceding example illustrates that a counting problem may often be attacked in several ways. Note that in this case, the method of approach 2 was easier than the direct attack in approach 1.

Example 10.4 A mother of eight children wishes to send three to the store, have two do dishes, and will let the other three play. How many ways can she divide up the children to accomplish this?

The mother must select three of her eight children to go to the store, which she can do in $\binom{8}{3}$ ways, *and* then she must also appoint two of the remaining five children to do the dishes, which she can do in $\binom{5}{2}$ ways. She really doesn't have to designate which three are to play; they are the children left over; of course, she can select them in $\binom{3}{3} = 1$ way. By Theorem 8.3, we should multiply for this *and* situation, so we have for our answer $\binom{8}{3}\binom{5}{2} = 56 \cdot 10 = 560$. ‖

EXERCISES

10.1 How many possibilities are there for the membership of a committee of four people chosen from a group of twelve people?

10.2 How many possibilities are there for the membership of a committee of four people chosen from a group of six men and six women if the committee is to contain two men and two women?

10.3 How many possibilities are there for the membership of a committee of four people chosen from a group of six men and six women if the committee is to contain more men than women?

10.4 From a club of twelve people, an entertainment committee of four people and an executive committee of three people are to be chosen.

a) How many possibilities are there for the memberships of the committees if anyone can serve on both committees?

b) How many possibilities are there for the memberships of the committees if no one can serve on both committees?

10.5 A mother of eight children wishes to send three to the store, have two do the dishes, and will let the rest play.

a) How many ways can she divide up the children to accomplish this if one of the children, Billy, is too small to sent to the store? (He can do the dishes.)
b) How many ways can she divide up the children to accomplish this if Billy is too small to go to the store (he can do the dishes) and Sue is too clumsy to do the the dishes (she can go to the store)?

10.6 In how many ways can eight students be assigned to two triple rooms and one double room?

10.7 In how many distinguishable ways can eight different elements be put into three boxes of different sizes? (It is permissible to put all eight elements in one box.)

10.8 In how many distinguishable ways can eight different elements be put into three boxes of different sizes with two in the largest box and three in each of the others?

10.9 Show that the total number of ways that n different elements can be put into r different boxes is r^n, generalizing Exercise 10.7.

10.10 Show that the number of ways that n different elements can be put into r different boxes, with n_1 in the first box, n_2 in the second, etc., where of course $n_1 + n_2 + \cdots + n_r = n$, is

$$\frac{n!}{(n_1!)(n_2!) \cdots (n_r!)},$$

generalizing Exercise 10.8.

10.11 Use Theorems 9.1 and 10.1 to show, without any computation, that

$$2^n = \binom{n}{0} + \binom{n}{1} + \binom{n}{2} + \cdots + \binom{n}{n-1} + \binom{n}{n}.$$

10.12 How many distinguishable ways can a grade of A, B, C, D or F be assigned to each of four students in a small seminar if

a) there are no restrictions?
b) each student gets a different grade?
c) exactly two students get grades of A?
d) exactly two students get grades of A, and exactly three students pass? (The only failing grade is F.)

10.13 A mother of young triplets, one boy and two girls, has seven different sweaters (all the same size) to distribute among them. She wishes to give three sweaters to the boy and two to each of the girls.

a) How many ways can she divide the sweaters up to accomplish this?
b) How many ways can she divide the sweaters up to accomplish this if two of the sweaters are of a color unsuitable for a boy?

10.14 A five-question multiple-choice test has three possible answers, a, b, and c, for each question.

a) How many different ways can the test be answered?
b) How many ways can the test be answered with the letter c never being used?

c) How many ways can the test be anwered with exactly two questions answered with the letter b?

d) How many ways can the test be answered with exactly two questions answered with the letter b and exactly one answered with the letter c?

e) How many ways can the test be answered with two letters each used exactly twice and the other letter used once?

10.15 How many distinguishable ways can

a) six different colored balls be arranged in a row?

b) two white and four black balls all the same size be arranged in a row?

c) three black balls, two white balls, and one red ball of the same size be arranged in a row?

10.16 Show that if a finite set U has an odd number of elements, then U has the same number of subsets containing an even number of elements as subsets containing an odd number of elements. (Zero is considered to be an even number.) [*Hint:* Use an argument suggested by Fig. 10.2].

10.17 Show that the conclusion of Exercise 10.16 is also true if U is a finite set with an even number n of elements, where $n \geq 2$. [*Hint:* Paint one element of U red, so that you can talk about it. Then using Exercise 10.16, count how many subsets contain an even number of elements; each of these subsets either contains the red element *or* does not contain the red element.]

REFERENCES

1. C. B. ALLENDOERFER and C. O. OAKLEY, *Principles of Mathematics*, 2nd ed. New York: McGraw-Hill, 1963.

2. FLORA DINKINES, *Elementary Theory of Sets*. New York: Appleton-Century-Crofts, 1964.

3. H. EVES and C. V. NEWSOM, *An Introduction to the Foundations and Fundamental Concepts of Mathematics*, rev. ed. New York: Holt, Rinehart and Winston, 1965.

4. P. R. HALMOS, *Naive Set Theory*. Princeton, N.J.: Van Nostrand, 1960.

5. J. G. KEMENY, J. L. SNELL, and G. L. THOMPSON, *Introduction to Finite Mathematics*, 2nd ed. Englewood Cliffs, N.J.: Prentice-Hall, 1966.

2 | Probability

The study of probability goes back to 1654. In that year, Pierre de Fermat (1601–1665) and Blaise Pascal (1623–1662) exchanged several letters regarding a matter of chance proposed to Pascal by a gambler, Chevalier de Méré. The subject has attracted the attention of some of the best mathematical minds since its inception, with Joseph Louis Lagrange (1736–1813) and Pierre Simon Laplace (1749–1827) making major contributions. Probability theory is a very active branch of mathematics today.

Probability is concerned with chance and relative likelihood. This branch of mathematics and the related field of statistics are potentially useful in analyzing any situation whose outcome is uncertain. Important practical uses of statistics in matters ranging from insurance to quality control in manufacturing are well known. Statistics is concerned with the application of the mathematical theory of probability. To fully understand statistics, some knowledge of probability is necessary.

In addition to its more pedestrian uses, probability lies at the heart of many theoretical studies in physics, biology, genetics, psychology, economics, and other sciences. Perhaps its most impressive entry into scientific theories is in the field of physics, where it has become increasingly important in this century. Many physicists now feel that there is a large element of chance in the physical behavior of our universe, and that probabilistic models and theories are appropriate for the study of these questions. This opinion was not shared by Albert Einstein, who was reported to have said, "I cannot believe that God plays dice with the universe." Probabilistic models and theories are extremely important in the relatively new field of particle physics (quantum mechanics), as we shall attempt to indicate in Section 7.

Although probability must be regarded as one of the most important branches of mathematics in terms of its applications to other disciplines, it does not, at the moment, have much application to other branches of mathematics. One can be a fine geometer, algebraist, topologist, analyst, or mathematical logician, and know essentially no probability. However, one cannot study probability in depth without knowing a lot of analysis, and present-day analysis depends, in turn, on algebra, geometry, and topology. Thus an expert in probability has to know a lot. It is easy to give an intuitive approach to the subject; this we shall do in Sections 1 through 7. However, the derivation of the deeper material treated in Section 8 requires some analysis, and we shall not attempt to derive these results in this text.

1. THE NOTION OF PROBABILITY

1.1 Events

Probability is the mathematics of chance. Frequently one wishes to determine how likely it is that some *event* occurs (or has occurred, or will occur). This is precisely the question with which probability is concerned. We shall use *"e"* to denote an event under consideration. For example, *e* might be the event

> *A head appears when a fair coin is flipped,*

or

> *The sum of the numbers which appear on*
> *two rolls of a fair die is 6,*

or

> *The republican candidate will win in the*
> *next presidential election.*

The *probability of an event e* is a *number* which measures the likelihood that *e* occurs. We shall denote this number by "pr[*e*]". It seems reasonable to require that pr[*e*] have the following properties.

1) Two events e_1 and e_2 are equally likely if and only if pr[e_1] = pr[e_2]. That is, equally likely events should have the same probability, and events having the same probability should be equally likely.

2) An event e_1 is more likely to occur than an event e_2 if and only if pr[e_1] > pr[e_2]. That is, the more likely it is that an event occur, the larger its probability should be, and conversely. Also, probabilities should indicate likelihood ratios. For example, if e_1 is twice as likely to occur as e_2, we require that pr[e_1] = 2 · pr[e_2].

There are two extreme cases for the probability of an event *e*.

CASE 1. It may be impossible for *e* to occur. For example, if a gumdrop is drawn at random from a bag of six gumdrops, all of which are black, and if *e* is the event

> *A red gumdrop is obtained,*

then *e* can't occur. By properties (1) and (2) above, the probability of such an impossible event *e* should be the smallest number which could be the probability of an event. Mathematicians have agreed to take the number 0 as the probability of an impossible event.

CASE 2. It may be certain that event *e* occurs. For example, if a gumdrop is drawn at random from a bag of six gumdrops, all of which are black, and if *e* is the event

> *A black gumdrop is obtained,*

then *e* must occur. Such an event must have as probability the largest number which could be the probability of an event. Mathematicians have agreed to let 1 be the probability of an event which is sure to occur.

These extreme values 0 and 1 for probabilities are simply conventions, but they are useful conventions, as we shall see. Please remember, *if you are asked to find the probability of an event, your answer has to be a number in* [0, 1].

There is one more property which it seems reasonable to require of $\text{pr}[e]$. We first give an important definition.

Definition. Two events e_1 and e_2 are **disjoint** if it is impossible for them both to occur.

If e_1 and e_2 are events, we can form from them the new events

$$e_1 \text{ and } e_2,$$

denoted by "$e_1 \wedge e_2$", and

$$e_1 \text{ or } e_2 \text{ (or both)},$$

denoted by "$e_1 \vee e_2$". By our convention in Case 1 above, if events e_1 and e_2 are disjoint, then $\text{pr}[e_1 \wedge e_2] = 0$. We now state our third requirement.

3) If events e_1 and e_2 are disjoint events, then

$$\text{pr}[e_1 \vee e_2] = \text{pr}[e_1] + \text{pr}[e_2].$$

Example 1.1 Suppose a fair coin is flipped once. Consider the events

$$e_1: \quad A \text{ head appears,}$$

$$e_2: \quad A \text{ tail appears.}$$

Clearly, e_1 and e_2 are disjoint events; that is, it is impossible that both a head and a tail appear in one flip of a coin. Thus $\text{pr}[e_1 \wedge e_2] = 0$, and

$$\text{pr}[e_1 \vee e_2] = \text{pr}[e_1] + \text{pr}[e_2].$$

On the other hand, the event $e_1 \vee e_2$ is certain to happen, for one flip of a coin must produce either a head or a tail. Therefore $\text{pr}[e_1 \vee e_2] = 1$. Since our coin is fair, we are as likely to get a head as a tail, that is, $\text{pr}[e_1] = \text{pr}[e_2]$. Thus we have $\text{pr}[e_1 \vee e_2] = \text{pr}[e_1] + \text{pr}[e_2] = 1$ and in addition $\text{pr}[e_1] = \text{pr}[e_2]$. Obviously we must have

$$\text{pr}[e_1] = \text{pr}[e_2] = \tfrac{1}{2}. \; \|$$

Example 1.1 indicates why it is convenient to let 1 be the probability of an event which is sure to occur. We can view $\text{pr}[e_1] = \tfrac{1}{2}$ as signifying that if a fair coin is flipped over and over, a head will appear about *half* the time. Similarly, $\text{pr}[e_1 \vee e_2] = 1$ indicates that either a head or a tail will appear all the time, that is, the *whole* time.

For any event e, we can consider the **complementary event** $\bar{e}$ given by

$$\bar{e}: \quad \textit{The event } e \textit{ does not occur.}$$

Then $e \vee \bar{e}$ is the event

$$\textit{Either } e \textit{ occurs or } e \textit{ does not occur.}$$

Clearly, e and $\bar{e}$ are disjoint events and $e \vee \bar{e}$ is certain to occur. Thus $\text{pr}[e \vee \bar{e}] = \text{pr}[e] + \text{pr}[\bar{e}] = 1$, so

$$\text{pr}[\bar{e}] = 1 - \text{pr}[e].$$

Example 1.2 For the events e_1 and e_2 in Example 1.1, we have $e_2 = \overline{e_1}$, and $\text{pr}[e_2] = \text{pr}[\overline{e_1}] = 1 - \text{pr}[e_1] = 1 - \frac{1}{2} = \frac{1}{2}$. ‖

1.2 Sample Spaces

To determine the probability of an event e, one usually considers the set of all possible outcomes for the situation with which e is concerned. Each possible outcome can itself be considered to be an event, an *outcome event*. The event e may or may not be one of the outcome events. We illustrate these ideas with two examples.

Example 1.3 Let a fair coin be flipped once and consider the event

$$e: \quad \textit{A head appears.}$$

As possible outcomes for this experiment in which a coin is flipped once, it is natural to take as outcome events

$$e_1: \quad \textit{A head appears,}$$
$$e_2: \quad \textit{A tail appears.}$$

In this case, our event e is the same as the outcome e_1. ‖

Example 1.4 Let a fair die be rolled once and consider the event

$$e: \quad \textit{An odd number appears.}$$

For this situation in which a die is rolled once, it is natural to take as outcome events the numbers 1, 2, 3, 4, 5, and 6, describing the number of dots which come up on the top face of the die. Here our event e under consideration is not one of our outcome events 1, 2, 3, 4, 5, 6. ‖

With these examples to guide us, we make a definition.

Definition. A **sample space** (or **possibility space**) S **for an experiment** (**situation**) is a set of outcome events for the experiment such that any two distinct outcome events e_i and e_j in S are disjoint, and such that some outcome in S has to occur, no matter what happens.

We may think of a sample space for a situation as an exhaustive set of disjoint outcome events, where the term *exhaustive* means that every contingency is covered by some outcome event in the set. A sample space really plays the role of a universal set for the situation.

Example 1.5 Suppose a fair die is rolled once. The set $\{1, 2, 3, 4, 5, 6\}$ is a sample space for this experiment. The set $\{1, 2, 3, 4, 5\}$ is not a sample space; this set is not exhaustive since the possibility that a 6 is rolled is not covered by any of the outcomes 1, 2, 3, 4, or 5. The set $\{odd, 2, 4, 5, 6\}$ is also not a sample space, since the events *odd* and 5 are not disjoint, for 5 is an odd number. ∥

Definition. If e is an event related to an experiment, then a sample space S for the experiment is a **sample space for** e if for each outcome event in S, e either definitely occurs or definitely does not occur.

Example 1.6 Let a fair die be rolled once, and consider the event

$$e: \quad A \ number \ greater \ than \ 2 \ comes \ up.$$

While $\{even, odd\}$ is a sample space for the experiment, it is *not* a sample space for e, for if you just know that the event *odd* occurs, you cannot definitely say whether or not e occurs. ∥

In practice, given an event e, every sample space you will naturally consider will be a sample space for e.

In this text, we shall deal only with *finite* sample spaces. In more advanced discussions of probability, infinite sample spaces are also considered. We stated before that if e is an impossible event, then $\text{pr}[e] = 0$. For our *finite* sample spaces, it is also true that if $\text{pr}[e] = 0$, then e is an impossible event. Suppose, however, that we have some random way of choosing a single real number from the set $\mathbf{R}$, and consider the event

$$e: \quad The \ number \ \pi \ is \ chosen.$$

The set $\mathbf{R}$ can be taken as a sample space for e, and $\mathbf{R}$ is an infinite set. It is surely *possible* for e to occur, but a little thought will convince you that we could not have $\text{pr}[e] > 0$ and also have our properties (1), (2), and (3) for probability hold. In the probability theory of infinite sample spaces, for this event e one has $\text{pr}[e] = 0$, even though e is a possible event.

From now on, we shall assume that all our sample spaces are finite. For finite sample spaces, $\text{pr}[e] = 0$ if and only if e is an impossible event, and similarly, $\text{pr}[e] = 1$ if and only if e is certain to occur.

1.3 Probability Measures

The first step in determining the probability of an event e is to find the probability of each outcome event e_i in a sample space S for e. Since these outcome events are disjoint and one of them must occur, the sum of their probabilities should be 1. Let us consider the probability of an outcome event to be the *weight of the outcome*. The next definition is a natural extension of these ideas.

> **Definition.** Let S be a finite set. A ***probability measure on*** S is an assignment of a nonnegative real number as weight to each element of S, subject to the requirement that the sum of all the weights is 1.

To find the probability of an event e, we first assign a measure to a sample space S for e such that the weight assigned to an outcome event $e_i \in S$ can appropriately be considered to be $\text{pr}[e_i]$. In real life situations, this is not always easy to do; as we illustrate in Example 1.9, it may require a statistical study. For the simple examples with which we shall deal, the assignment of weights to outcomes in a sample space usually causes no problem.

In Section 2 we shall define $\text{pr}[e]$ in terms of a probability measure on a sample space for e. First, we want you to think about the concepts we have introduced, and get an intuitive feeling for what $\text{pr}[e]$ means. The exercises are designed to help you accomplish this. We conclude with three more examples.

Example 1.7 Consider the experiment in which a fair die is rolled once, as in Example 1.4. We have as sample space the set $\{1, 2, 3, 4, 5, 6\}$, Since the die is fair, these outcomes are equally likely to occur, so they should have equal weight (probability). Since the sum of these weights must be 1, each should have a weight $\frac{1}{6}$. This gives the desired probability measure on $\{1, 2, 3, 4, 5, 6\}$ for this experiment. ‖

Example 1.8 Suppose a coin is biased in such a way that a tail is twice as likely to come up as a head when the coin is flipped. (This might be determined by making a statistical study of what happens when the coin is flipped repeatedly.) For the experiment consisting of a single flip of this coin, we can take as sample space the set $\{H, T\}$, where H and T have the obvious meanings. If the weight (probability) of H is x, then the weight of T should be $2x$, since a tail is twice as likely to come up as a head. For a probability measure, we must have $x + 2x = 1$ or $3x = 1$. Thus $x = \frac{1}{3}$, so $\text{pr}[H] = \frac{1}{3}$ and $\text{pr}[T] = \frac{2}{3}$. ‖

Example 1.9 Suppose an election between candidates A and B is to take place. We may take as sample space the set $\{A \text{ wins}, B \text{ wins}, \text{tie}\}$. *Without further information, it is impossible to say what the appropriate probability measure might be for this space.* The purpose of public opinion polls is to try to estimate reasonable weights for such outcomes. If you knew that everyone would vote at random, say by flipping a fair coin, and that an odd number of people would vote, so that a tie would be impossible, then you would know that $\text{pr}[A \ wins] = \frac{1}{2}$, $\text{pr}[B \ wins] = \frac{1}{2}$, and $\text{pr}[tie] = 0$. Without such information, you just can't say. Never assume more than you are actually given. Some of the exercises that follow emphasize this point. ‖

EXERCISES

1.1 Given that e is an event such that $\text{pr}[e] = \frac{1}{3}$, find the following.

a) $\text{pr}[\bar{e}]$ b) $\text{pr}[e \wedge \bar{e}]$ c) $\text{pr}[e \vee \bar{e}]$ d) $\text{pr}[e \wedge e]$
e) $\text{pr}[e \vee e]$ f) $\text{pr}[\bar{e} \wedge \bar{e}]$ g) $\text{pr}[\bar{e} \vee \bar{e}]$

1.2 Suppose we are concerned with the calendar day of the year on which Bill's birthday falls.

a) How many elements are there in the natural sample space for this situation?
b) Would it be appropriate to assign each element in the natural sample space the same weight, in a probability measure?
c) Suppose you know that Bill was not born on February 29. With this information, how many elements are there in the natural sample space for this situation? Do you think it would now be appropriate to assign each outcome equal weight in a probability measure?

1.3 Suppose a man has two quarters, three dimes, one penny, and one nickel in his pocket, and selects a coin from his pocket. From the given data, is it possible to find an appropriate probability measure for the sample space $\{Q, D, P, N\}$?

1.4 With reference to Exercise 1.3, assuming that the man selects the coin from his pocket by some random method, find the appropriate probability measure on the sample space $\{Q, D, P, N\}$.

1.5 Let a fair coin be flipped twice. Which of the following are sample spaces for this experiment? What is wrong with the others? The abbreviated outcomes have the obvious meanings.

a) $\{2H, 2T\}$
b) $\{no \ H, \ exactly \ one \ H, \ exactly \ two \ H\}$
c) $\{more \ H \ than \ T, \ more \ T \ than \ H\}$
d) $\{HH, HT, TH, TT\}$
e) $\{at \ least \ one \ H, \ at \ least \ one \ T\}$

1.6 Let a fair die be rolled once. Which of the following are sample spaces for this experiment? What is wrong with the others? The abbreviated outcomes have the obvious meanings.

a) $\{>3, <3\}$ b) $\{1, 2, 3, >3\}$ c) $\{1, 2, \geq 2\}$
d) $\{odd, even\}$ e) $\{\leq 2, 3, 4, >4\}$

1.7 A fair die is rolled once. Consider the event

$$e: \quad A \ number \ greater \ than \ 4 \ comes \ up.$$

Which of the following are sample spaces for e? What is wrong with the others?

a) $\{1, 2, 3, 4, 5, 6\}$ b) $\{\leq 3, 4, 5, 6\}$
c) $\{1, 3, 5, even\}$ d) $\{<5, \geq 5\}$
e) $\{1 \ or \ 4, \ 2 \ or \ 5, \ 3 \ or \ 6\}$

1.8 A coin is biased so that a head is four times as likely to come up as a tail when the coin is flipped. The coin is flipped once. Determine the appropriate probability measure on the sample space $\{H, T\}$.

1.9 A die is loaded in such a way that any odd number is as likely to come up as any other odd number, but each even number is twice as likely to come up as any odd number. The die is rolled once. Determine the appropriate probability measure on the sample space $\{1, 2, 3, 4, 5, 6\}$.

1.10 A die is loaded in such a way that the likelihood that a face comes up is proportional to the number of dots on the face. That is, a 6 is three times as likely to come up as a 2, etc. The die is rolled once. Determine the appropriate probability measure on the sample space $\{1, 2, 3, 4, 5, 6\}$.

1.11 A bag contains three red gumdrops, two green ones, and one yellow gumdrop. A gumdrop is drawn at random from the bag. Give the appropriate probability measure on each of the following sample spaces. The abbreviations have the obvious meanings.

a) $\{R, G, Y\}$ b) $\{R_1, R_2, R_3, G_1, G_2, Y\}$
c) $\{R, G_1, G_2, Y\}$ d) $\{R_1, G, other\}$

1.12 Let one card be chosen at random from a deck of 52 playing cards. Determine the probability measure on each of the given sample spaces.

a) $\{black, red\}$ b) $\{black, heart, diamond\}$
c) $\{club, space, heart, diamond\}$ d) $\{jack, queen or king, ace, other\}$

1.13 If there is an "even chance" that an event e occurs, what is $\mathrm{pr}[e]$?

1.14 If the "odds that an event e occurs are 2 to 1," what is $\mathrm{pr}[e]$?

1.15 If the "odds that an event e occurs are 3 to 2," what is $\mathrm{pr}[e]$?

1.16 If the "odds that an event e occurs are r to s," what is $\mathrm{pr}[e]$?

1.17 Consider a set composed of n people. Let e be the event

> *At least two of the people have their birthdays*
> *on the same calendar day of the year.*

If $n = 2$ and the people were selected at random, clearly e is not likely to occur, and $\mathrm{pr}[e]$ is close to 0. If $n = 367$, then e is certain to occur, since there are at most 366 different calendar days in any year; in this case, $\mathrm{pr}[e] = 1$. There should be a value of n somewhere between 2 and 367 for which there is about an even chance that e occurs, so that $\mathrm{pr}[e]$ for this n would be about $\frac{1}{2}$.

a) Think about this problem for a minute, and then *guess* for what value of n we should have $\mathrm{pr}[e]$ close to $\frac{1}{2}$.

b) Look up the surprising answer to this problem in the back of the text. This problem is discussed in Kemeny, Snell, and Thompson [3], where a table is given which shows $\mathrm{pr}[e]$ for many values of n and which gives the odds that e occurs. The values for $\mathrm{pr}[e]$ given in that table are probably a trifle too low, since they were worked out on the assumption that a person's birthday is as likely to fall on one calendar day of the year as on another (see Exercise 1.2). Discover whether there are two people in your class whose birthdays are on the same calendar day.

2. PROBABILITY

2.1 The Probability of an Event

Let S be a finite set and let a probability measure be given on S. For example, S might be the sample space of an event e with the appropriate probability measure. If A is a subset of S, we shall let the **measure $m(A)$ of A** be the sum of the weights of all the elements of A. Clearly, we always have

$$0 \leq m(A) \leq 1.$$

If S is a sample space for an event e, the **event set E of e** is the set of all outcomes in S for which e occurs.

Example 2.1 Let a fair die be rolled once, and consider the event

$$e: \quad An\ odd\ number\ appears.$$

For the sample space $\{1, 2, 3, 4, 5, 6\}$, the appropriate probability measure assigns a weight of $\frac{1}{6}$ to each outcome. The event set E is $\{1, 3, 5\}$, for the outcomes 1, 3, and 5 are those for which e occurs. Note that

$$m(E) = \tfrac{1}{6} + \tfrac{1}{6} + \tfrac{1}{6} = \tfrac{1}{2}. \ \|$$

We shall now define the probability of an event e having a finite sample space. This definition should seem natural, in view of the preceding example.

Definition. Let e be an event and let S be a sample space for e with an appropriate probability measure. Let $E \subseteq S$ be the event set of e. Then the **probability** $\mathrm{pr}[e]$ **of e** is $m(E)$.

Example 2.2 We see from Example 2.1 that the probability is $\frac{1}{2}$ that an odd number comes up when a fair die is rolled once. Of course, this is intuitively obvious. $\|$

We have defined $\mathrm{pr}[e]$ in terms of sets. Let e_1 and e_2 be events with the same sample space and with event sets E_1 and E_2, respectively. It follows at once from the preceding definition and from the definitions of intersection and union of sets that

$$\mathrm{pr}[e_1 \wedge e_2] = m(E_1 \cap E_2)$$

and

$$\mathrm{pr}[e_1 \vee e_2] = m(E_1 \cup E_2).$$

This suggests that Venn diagrams may be useful for solving some probability problems. The technique is best illustrated with an example.

Example 2.3 A weatherman predicts that there will be some precipitation (either rain or snow) the next day with probability $\frac{5}{6}$, that it will rain with probability $\frac{1}{2}$, and that it will snow with probability $\frac{3}{4}$. Let us find his predicted probability that it will both rain and snow the next day.

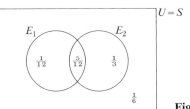

Fig. 2.1

Let us take for events

e_1:   *It will rain the next day,*

e_2:   *It will snow the next day.*

We may regard the event sets E_1 and E_2 as subsets of some sample space which we take as universal set. We draw a two-subset Venn diagram, and label each of the four regions with the appropriate measure of the subset of S which it represents, as shown in Fig. 2.1. This is done as follows. Since it either rains or snows (or both) with probability $\frac{5}{6}$, we have $m(E_1 \cup E_2) = \frac{5}{6}$, so $m(\overline{E_1 \cup E_2}) = \frac{1}{6}$, which we fill in on the diagram in Fig. 2.1. Since $m(E_1 \cup E_2) = \frac{5}{6}$ and $m(E_1) = \frac{1}{2}$, we find that $m(\overline{E_1} \cap E_2) = \frac{5}{6} - \frac{1}{2} = \frac{1}{3}$. Similarly, $m(E_1 \cap \overline{E_2}) = \frac{5}{6} - \frac{3}{4} = \frac{1}{12}$. Thus we must have $m(E_1 \cap E_2) = \frac{5}{12}$ in order for the measures of the four regions to add up to one. Thus the probability of both rain and snow the next day is given by

$$\mathrm{pr}[e_1 \wedge e_2] = m(E_1 \cap E_2) = \tfrac{5}{12}. \;\|$$

An event e is completely determined by its event set E in a sample space. We told you in Chapter 1 that mathematicians attempt to base all their work on the concept of a set. We did not define an *event* in the last section. In more advanced treatments of probability, an event is considered to be the same as our *event set*. That is, one defines a (finite) **sample space** to be a (finite) set S with a probability measure, and one then defines an **event** E to be a subset of S. *Thus in more advanced presentations an event is defined to be a set.*

2.2 The Equiprobable Measure

To find the probability of an event e, we usually proceed as follows.

STEP 1. Find a sample space S for e.

STEP 2. Assign an appropriate probability measure to S.

STEP 3. Find the event set E of e.

STEP 4. Compute $m(E) = \mathrm{pr}[e]$.

Frequently in Step 2 one finds that the outcomes in S are all equally likely.

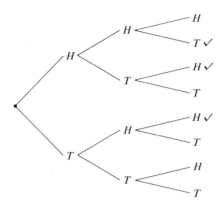

Fig. 2.2

If this is the case, then the weight assigned to each outcome should be $1/n(S)$, where $n(S)$ is the number of elements in S. This measure is the ***equiprobable measure on*** S. The measure of the event set E is then just

$$n(E) \cdot \frac{1}{n(S)} = \frac{n(E)}{n(S)},$$

where $n(E)$ is the number of outcomes in E. We state this result as a theorem, and conclude with some examples.

Theorem 2.1 *If the equiprobable measure is appropriate for a sample space S of an event e, then*

$$\mathrm{pr}[e] = \frac{n(E)}{n(S)}.$$

Example 2.4 Let a fair coin be flipped three times in succession. Let us find the probability that exactly two heads appear.

We let e be the event

Exactly two heads appear.

The tree diagram in Fig. 2.2 is a convenient way for us to represent a sample space S for e. The tree has eight terminal points, so $n(S) = 8$. Since the coin is fair, a head is as likely to appear as a tail on each flip. Thus we are as likely to follow any one path of our tree as any other path, so the equiprobable measure is appropriate. The path outcomes corresponding to outcomes in E are checked in Fig. 2.2. We have $n(E) = 3$. Therefore

$$\mathrm{pr}[e] = \frac{n(E)}{n(S)} = \frac{3}{8}. \parallel$$

Example 2.5 Let a fair die be rolled twice, and consider the event

e: *The sum of the numbers which come up is 5.*

Let us find $\mathrm{pr}[e]$.

Any one of six numbers can appear on the first roll of our die, *and* any one of six on the second roll, so by our counting arguments in Chapter 1, any one of $6 \cdot 6 = 36$ different sequences of two rolls can occur. Since our die is fair, any one of the 36 outcomes is as likely to happen as any other, so the equiprobable measure is appropriate. A typical outcome can be denoted by the ordered pair of numbers formed from the numbers which appear on the first and second rolls of the die. For example, (2, 4) is the outcome in which a 2 appears on the first roll and a 4 on the second. We have

$$E = \{(1, 4), (2, 3), (3, 2), (4, 1)\},$$

so

$$\text{pr}[e] = \frac{n(E)}{n(S)} = \frac{4}{36} = \frac{1}{9}. \parallel$$

Example 2.6 A *bridge hand* consists of a 13-card subset of the set of 52 cards which comprises an ordinary deck of playing cards. Let us find the probability that a bridge hand chosen at random contains all four kings.

We let *e* be the event

The hand contains all four kings.

We can take as sample space *S* the set of all possible bridge hands. Since each hand consists of 13 cards chosen from the 52, our work in Chapter 1 shows that $n(S) = \binom{52}{13}$. Since the cards are chosen at random, one hand is as likely as any other, and the equiprobable measure is appropriate. It remains for us to count the number of elements in the event set *E*. To form a hand in *E* containing all four kings, we must select the four kings, which can be done in $\binom{4}{4} = 1$ way, *and* then select 9 cards from the remaining 48 to complete our bridge hand. Thus $n(E) = 1 \cdot \binom{48}{9}$, and

$$\text{pr}[e] = \frac{n(E)}{n(S)} = \frac{\binom{48}{9}}{\binom{52}{13}}.$$

This is a perfectly acceptable form for the answer. A bit of dull work with a pencil shows that

$$\text{pr}[e] = \frac{13 \cdot 12 \cdot 11 \cdot 10}{52 \cdot 51 \cdot 50 \cdot 49} = \frac{11}{17 \cdot 5 \cdot 49},$$

which is approximately 0.0026. It is not very likely to happen. $\parallel$

EXERCISES

2.1 Find the probability that three flips of a fair coin produce exactly one head.

2.2 A die is loaded in such a way that any odd number is as likely to come up as any other odd number, but each even number is twice as likely to come up as any odd number. The die is rolled once. Find the probability that a number greater than 3 comes up.

2.3 Judging from Bill's past history, his friends predict that he will date either a blonde or a brunette (or both) next week with probability $\frac{1}{2}$, and that he will date a blonde with probability $\frac{1}{3}$. What do they feel is the probability that he will date a brunette next week, but not a blonde?

2.4 Each of three bags contains one red gumdrop, one yellow one, and one green one. Three gumdrops are drawn at random, one from each bag.

a) Find the probability that the gumdrops are all green.
b) Find the probability that the gumdrops are all the same color.
c) Find the probability that exactly one gumdrop is red.
d) Find the probability that one gumdrop of each color is drawn.

2.5 Find the probability that the sum of the numbers which come up on two rolls of a fair die is 6.

2.6 If a fair die is rolled twice, what value is the most likely for the sum of the two numbers which come up?

2.7 Find the probability of obtaining more heads than tails if a fair coin is flipped

a) once,   b) twice,
c) three times,   d) four times.

2.8 Argue by symmetry that if a fair coin is flipped an *odd* number of times, the probability of obtaining more heads than tails is $\frac{1}{2}$.

2.9 Argue by symmetry that if a fair coin is flipped an *even* number $2n$ times, the probability of obtaining more heads than tails is

$$\frac{1 - \dfrac{\binom{2n}{n}}{2^{2n}}}{2}.$$

[*Hint:* Consider the sample space

{*more H than T, more T than H, same number of H as T*}.]

2.10 Find the probability that a random bridge hand has exactly five hearts. (Don't work out your answer.)

2.11 Find the probability that a random bridge hand has exactly five hearts and exactly three spades. (Don't work out your answer.)

2.12 Find the probability that a random bridge hand has exactly five spades and exactly four red cards. (Don't work out your answer.)

2.13 A poker hand is a five-card subset of the set of 52 playing cards. Find the probability that a random poker hand has four cards of the same face value.

2.14 Three boys and two girls are placed at random in a row for a picture. Find the probability that the girls stand next to each other.

2.15 Show that for any events e_1 and e_2, we have

$$\text{pr}[e_1 \lor e_2] = \text{pr}[e_1] + \text{pr}[e_2] - \text{pr}[e_1 \land e_2].$$

[*Hint:* This is analogous to our formula $n(A \cup B) = n(A) + n(B) - n(A \cap B)$ in Chapter 1. Draw a Venn diagram, and argue that $m(E_1 \cup E_2) = m(E_1) + m(E_2) - m(E_1 \cap E_2)$.]

3. CONDITIONAL PROBABILITY

3.1 Conditional Probability

Sometimes you want to find the probability of an event e_1 when you know that another event e_2 with the same sample space as e_1 does occur. The probability that e_1 occurs, given that e_2 occurs, is the **conditional probability of e_1, given e_2**, and is denoted by "$\mathrm{pr}[e_1 \mid e_2]$".

Let e_1 be an event with sample space S, and let e_2 be an event with the same sample space. If you are *sure* that e_2 occurs, you are sure that only the outcome events in E_2 can occur. Often the easiest way to compute $\mathrm{pr}[e_1 \mid e_2]$ is to think of E_2 as a new sample space for e_1, and to assign an appropriate probability measure to E_2. Of course, you must assign the measure so that the total measure of E_2 is 1. Alternatively, one can keep the original sample space S for e_1, but assign a new probability measure on S so that an outcome not in E_2 has a weight of 0. We illustrate these ideas with some examples.

Example 3.1 Let a fair coin be flipped three times, and let us find the conditional probability that a tail appears on the second flip, given that at least one head appears.

We take as events

$$e_1: \quad A \text{ tail appears on the second flip,}$$
$$e_2: \quad At \text{ least one head appears.}$$

We want to find $\mathrm{pr}[e_1 \mid e_2]$. Figure 3.1 shows the tree diagram for three flips of a coin. The eight possible outcomes are equally likely. Now we know that e_2 does occur. Each of the seven outcomes in the event set E_2 is marked with a $\times$ in Fig. 3.1. Since they are equally likely, if we take E_2 as our new sample space, each should have a weight of $\frac{1}{7}$. Those outcomes in E_2 for which e_1 occurs are checked in our figure. Since there are three of them, we have $\mathrm{pr}[e_1 \mid e_2] = 3 \cdot \frac{1}{7} = \frac{3}{7}$. *Note that the bottom path of our tree is not checked,* for even though e_1 occurs for this outcome, the outcome is not in our event set E_2. ‖

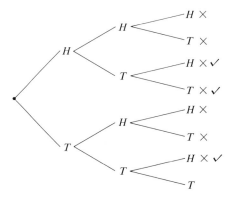

Fig. 3.1

Example 3.2 Let us find the probability that a random bridge hand has exactly four hearts, given that it has exactly six red cards.

We take as events

e_1:   *The hand has exactly four hearts,*

e_2:   *The hand has exactly six red cards.*

Since the hand is chosen at random, the equiprobable measure on the new sample space E_2 is appropriate. A bridge hand with exactly six red cards must also have exactly seven black cards, and there are $\binom{26}{6} \cdot \binom{26}{7}$ such hands. We are interested in the number of these hands in E_2 which have exactly four hearts. Such a hand must have four hearts, two diamonds (to make up six red cards), and seven black cards, so there are $\binom{13}{4} \cdot \binom{13}{2} \cdot \binom{26}{7}$ such hands. Thus

$$\text{pr}[e_1 \mid e_2] = \frac{\binom{13}{4} \cdot \binom{13}{2} \cdot \binom{26}{7}}{\binom{26}{6} \cdot \binom{26}{7}} = \frac{\binom{13}{4} \cdot \binom{13}{2}}{\binom{26}{6}}. \; \|$$

The conditional probability $\text{pr}[e_1 \mid e_2]$ can also be computed as a quotient of two ordinary probabilities, using a formula which we now develop. We shall have use for this formula later. However, in computing the conditional probability $\text{pr}[e_1 \mid e_2]$, we personally prefer to regard E_2 as a new sample space, as we did in Examples 3.1 and 3.2.

When computing $\text{pr}[e_1 \mid e_2]$, you know that e_2 occurs. Thus e_2 is a possible event, so $\text{pr}[e_2] = m(E_2) > 0$. You know that one of the outcomes in the event set E_2 must occur. Knowing that you are in E_2, you want to determine how likely it is that you are also in E_1. The part of E_2 which is also in E_1 is of course $E_1 \cap E_2$. If S is a sample space for both e_1 and e_2, then the proportion of E_2 which is also in E_1 is given by

$$\frac{m(E_1 \cap E_2)}{m(E_2)},$$

as indicated in Fig. 3.2. We have arrived at the formula

$$\text{pr}[e_1 \mid e_2] = \frac{m(E_1 \cap E_2)}{m(E_2)} = \frac{\text{pr}[e_1 \wedge e_2]}{\text{pr}[e_2]}.$$

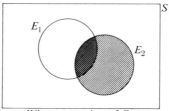

What proportion of E_2
is the crosshatched region?   **Fig. 3.2**

Example 3.3 Let us solve the problem posed in Example 3.1 again, using the formula

$$\text{pr}[e_1 \mid e_2] = \frac{\text{pr}[e_1 \wedge e_2]}{\text{pr}[e_2]} .$$

This amounts to taking a quotient of two probabilities. From the sample space S given by the whole tree in Fig. 3.1, with the equiprobable measure, we find that

$$\text{pr}[e_2] = \tfrac{7}{8}.$$

The event set $E_1 \cap E_2$ of $e_1 \wedge e_2$ (*a tail appears on the second flip and at least one head appears*) contains those outcomes checked in Fig. 3.1. Thus $\text{pr}[e_1 \wedge e_2] = \tfrac{3}{8}$. By our formula, we obtain

$$\text{pr}[e_1 \mid e_2] = \frac{\tfrac{3}{8}}{\tfrac{7}{8}} = \frac{3}{8} \cdot \frac{8}{7} = \frac{3}{7} .$$

Essentially the same cases had to be considered in this computation using the formula as were considered in Example 3.1, so we have not saved ourselves any work by using this formula. ‖

3.2 Independent Events

Let e_1 and e_2 be two possible events having the same sample space. Then neither $\text{pr}[e_1]$ nor $\text{pr}[e_2]$ is zero. The likelihood that e_1 occurs may or may not be affected by whether or not e_2 occurs. If the occurrence of e_2 has no effect on the probability of e_1, this means that $\text{pr}[e_1 \mid e_2] = \text{pr}[e_1]$; that is, knowing that e_2 occurs does not change $\text{pr}[e_1]$. Of course, it is also possible that the occurrence of e_2 may affect the likelihood of e_1, so that $\text{pr}[e_1 \mid e_2] \neq \text{pr}[e_1]$.

If $\text{pr}[e_1 \mid e_2] = \text{pr}[e_1]$ so that e_2 does not affect the likelihood of e_1, then e_1 is *independent of* e_2. Using our formula for conditional probability, we find that e_1 is independent of e_2 if and only if

$$\text{pr}[e_1 \mid e_2] = \frac{\text{pr}[e_1 \wedge e_2]}{\text{pr}[e_2]} = \text{pr}[e_1],$$

or if and only if

$$\text{pr}[e_1 \wedge e_2] = \text{pr}[e_1] \cdot \text{pr}[e_2].$$

From the symmetry of this last equation in e_1 and e_2, we see that e_1 is independent of e_2 if and only if e_2 is independent of e_1. That is, if e_2 does not affect the likelihood of e_1, then e_1 does not affect the likelihood of e_2 either. We take the last equation as a formal definition of independent events.

Definition. Two possible events e_1 and e_2 with the same sample space are *independent* if and only if $\text{pr}[e_1 \wedge e_2] = \text{pr}[e_1] \cdot \text{pr}[e_2]$.

Example 3.4 For the events e_1 and e_2 in Example 3.1, where $\mathrm{pr}[e_1]$ is the probability that a tail appears on the second flip of a fair coin, we obviously have $\mathrm{pr}[e_1] = \frac{1}{2}$. We found that $\mathrm{pr}[e_1 \mid e_2] = \frac{3}{7}$. Thus $\mathrm{pr}[e_1 \mid e_2] \neq \mathrm{pr}[e_1]$, so e_1 and e_2 are not independent events. We could also verify that the conditions of our definition do not hold, for it is easy to see that $\mathrm{pr}[e_1] = \frac{1}{2}$, $\mathrm{pr}[e_2] = \frac{7}{8}$, and $\mathrm{pr}[e_1 \wedge e_2] = \frac{3}{8}$. Thus

$$\mathrm{pr}[e_1 \wedge e_2] = \tfrac{3}{8} \neq \tfrac{1}{2} \cdot \tfrac{7}{8} = \mathrm{pr}[e_1] \cdot \mathrm{pr}[e_2]. \; \|$$

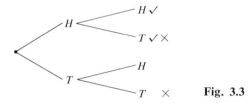

Fig. 3.3

Example 3.5 Let a fair coin be flipped twice, and consider the events

e_1: *A head appears on the first flip,*

e_2: *A tail appears on the second flip.*

Obviously the occurrence of either event should not affect the likelihood of the other, so the events should be independent. Let us verify that the condition in our definition holds. The tree representation of the natural sample space is given in Fig. 3.3. The equiprobable measure is appropriate. The outcomes in E_1 are checked and those in E_2 are marked with a ×. We have $\mathrm{pr}[e_1] = \frac{1}{2}$, $\mathrm{pr}[e_2] = \frac{1}{2}$, and $\mathrm{pr}[e_1 \wedge e_2] = m(E_1 \cap E_2) = \frac{1}{4}$. Thus

$$\mathrm{pr}[e_1 \wedge e_2] = \tfrac{1}{4} = \tfrac{1}{2} \cdot \tfrac{1}{2} = \mathrm{pr}[e_1] \cdot \mathrm{pr}[e_2]. \; \|$$

In the preceding example, it was intuitively obvious that our events e_1 and e_2 were independent. However, it is not always this obvious whether or not events are independent.

EXERCISES

3.1 A fair die is rolled once. Find the probability that a number greater than 3 comes up, given that an even number comes up.

3.2 A fair coin is flipped twice. Find the probability that both flips are heads, given that at least one flip is a head.

3.3 Find the probability that a head appears on the 20th flip of a fair coin, given that the first 19 flips were all tails.

3.4 A fair die is rolled twice. Find the probability that the sum of the numbers which come up is 6, given that at least one 4 comes up.

3.5 A fair die is rolled twice. Find the probability that the sum of the numbers which come up is 5, given that the sum is greater than 3.

3.6 A die is loaded in such a way that the probability that a number comes up is proportional to the number. That is, a 6 is three times as likely to come up as a 2, etc. If the die is rolled once, find the probability that an odd number comes up, given that a number less than 5 comes up.

3.7 Find the probability that a random bridge hand has exactly four spades, given that it has exactly six red cards.

3.8 Find the probability that a random bridge hand has exactly seven spades, given that it has exactly four black cards.

3.9 Each of three bags contains one red gumdrop, one yellow one, and one green one. One gumdrop is drawn at random from each bag.

a) Find the probability that the three gumdrops are all the same color, given that two of them are red.

b) Find the probability that the gumdrops are all the same color, given that none of them is red.

3.10 Each of three bags contains one red gumdrop, one yellow one, and one green one. One gumdrop is drawn at random from each bag. Determine whether the events in each pair of events are independent.

a) The gumdrops are all the same color; the first gumdrop drawn is red.

b) The gumdrops are all the same color; one of the gumdrops is red.

c) The gumdrops are all the same color; no gumdrop is red.

d) One gumdrop of each color is obtained; at least one red gumdrop is obtained.

3.11 A positive integer from 1 to 12 inclusive is selected at random. For each given pair of events, determine whether the events are independent.

a) An odd number is drawn; a number greater than 6 is drawn.

b) An odd number is drawn; a number greater than 5 is drawn.

3.12 Consider three circular cardboard disks of the same size. One disk is black on both sides, another is white on both sides, while the third is white on one side and black on the other. One of these disks is selected at random and placed on a table while your back is turned. You turn around and see that the top of the disk is white. If you guess that the bottom is also white, what is the probability that you are correct?

4. TREE MEASURES

4.1 Stochastic Processes and Trees

Many situations arise in probability for which it is convenient to represent a sample space for an event by drawing a tree. It is natural to draw a tree for an experiment which consists of a sequence of individual experiments (stages), performed in succession. For example, the experiment consisting of four flips of a coin can be broken down into a sequence of four individual experiments (stages), each consisting of a single flip of the coin. A finite sequence of experiments, where the outcome of each individual experiment is determined by chance, is a *finite stochastic process*. It is natural to represent a sample space for a stochastic process by a tree.

A typical tree is shown in Fig. 4.1. This tree corresponds to a two-stage experiment with possible outcomes a and b at the first stage, and possible outcomes x, y, and z at the second stage. Let us suppose that the probability of outcome a for the first stage is $\frac{1}{3}$. We label the branch connecting the starting point to the outcome a with the **branch weight** $\frac{1}{3}$. Since either a or b must occur, the appropriate branch weight leading to b is $\frac{2}{3}$. Let us also suppose that the outcomes x, y, and z for the second stage occur with probabilities $\frac{1}{2}$, $\frac{1}{4}$, and $\frac{1}{4}$, respectively. We have labeled the second stage branches with these branch weights in Fig. 4.1. An assignment of a nonnegative real number as weight to each branch of a tree in this manner, *so that the sum of the weights of all the branches which emanate from a single point is* 1, is a *tree measure*.

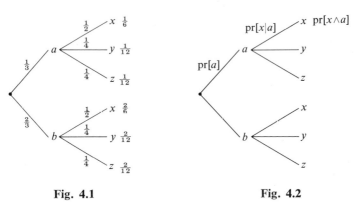

Fig. 4.1   Fig. 4.2

Our problem is to determine, from these branch weights, the weight of the path outcomes which form our sample space. We claim that to find the weight of a path outcome, we should *multiply* the branch weights on that path. Thus, referring to Fig. 4.1, the top path outcome has weight $\frac{1}{3} \cdot \frac{1}{2} = \frac{1}{6}$, etc. Let us now demonstrate that it is indeed appropriate to multiply the branch weights in order to find the weight of a path outcome.

We draw our tree again in Fig. 4.2, but without the hypothetical branch weights of Fig. 4.1. The branch weight which must be assigned to the branch leading to a is of course $\mathrm{pr}[a]$, and we so label this branch in Fig. 4.2. The branch leading from a to x on the top path should *not* be labeled $\mathrm{pr}[x]$, for the event set of outcome x also includes the fourth path from the top in our tree. The weight of this branch connecting a to x is actually *the probability that x occurs, given that a occurred*. Thus the appropriate weight for this branch is $\mathrm{pr}[x \mid a]$, and we so label this branch in Fig. 4.2. The top path may be described as the outcome in which both a and x occur, so its weight should be pr[$a \wedge x$], or, equivalently, pr[$x \wedge a$]. From our conditional probability formula of the preceding section, we have

$$\text{pr}[x \mid a] = \frac{\text{pr}[x \wedge a]}{\text{pr}[a]},$$

so

$$\text{pr}[x \wedge a] = \text{pr}[a] \cdot \text{pr}[x \mid a].$$

This shows that the **path weight** pr[$x \wedge a$] is indeed the product of our two branch weights pr[a] and pr[$x \mid a$]. Similar reasoning holds with any tree measure.

4.2 Some Illustrations

We give a few examples involving tree measures.

Example 4.1 A fair coin is flipped twice. Let us find the probability of the event

> *e: At least one head and at least one tail appear.*

The tree is shown in Fig. 4.3. Since the coin is fair, all branch weights should be $\frac{1}{2}$ as shown in the figure. The checked paths represent the outcomes in the event set E. Each has weight $\frac{1}{2} \cdot \frac{1}{2} = \frac{1}{4}$. Thus we have pr[$e$] $= \frac{1}{4} + \frac{1}{4} = \frac{1}{2}$. Of course, we could have solved this problem in Section 2. ‖

Example 4.2 A bag contains two red gumdrops, one yellow one, and one green one. Gumdrops are drawn successively at random without replacement until a red one is obtained, and then no more are drawn. Let us find the probability of the event

> *e: The green gumdrop is drawn.*

We are only interested in the color of the gumdrops drawn, so the tree in Fig. 4.4 represents a satisfactory sample space for e. Since for the first draw

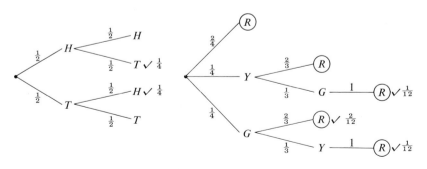

Fig. 4.3   **Fig. 4.4**

there are two red gumdrops but only one yellow gumdrop and one green one in the bag, our first stage branch weights should be $\frac{2}{4}$, $\frac{1}{4}$, and $\frac{1}{4}$, as shown in Fig. 4.4. Turning to the second stage, if the red gumdrop was not drawn on the first draw, so that our experiment continues, we are drawing at random from three gumdrops, two of which are red. Thus branch weights of $\frac{2}{3}$ and $\frac{1}{3}$ are appropriate, as shown in the figure. Of course, a third draw must produce a red gumdrop. The path outcomes in the event set E are checked, and we compute the weight of each checked path by multiplying the branch weights. We find that

$$\text{pr}[e] = \tfrac{1}{12} + \tfrac{2}{12} + \tfrac{1}{12} = \tfrac{4}{12} = \tfrac{1}{3}. \;\|$$

Example 4.3 For the experiment given in Example 4.2, let us find the probability that the green gumdrop is drawn, given that the yellow one is drawn.

We take as events

e_1:   *The green gumdrop is drawn,*

e_2:   *The yellow gumdrop is drawn,*

and we wish to find the conditional probability $\text{pr}[e_1 \mid e_2]$. The tree is given again in Fig. 4.5. This time, we mark with a × the outcomes in the event set E_2, and find the path weights of each of these outcomes. We then assign a new measure to E_2, so that the sum of the new weights of outcome events in E_2 is 1. Since we must preserve the relative likelihood of the outcomes in E_2, we see that we must assign double the weight to the top path we marked with a × that we assign to the other two. Thus our new weights must be of the form $2x$, x, and x, where $2x + x + x = 1$. We find that $x = \frac{1}{4}$, so our new weights are $\frac{2}{4}$, $\frac{1}{4}$, and $\frac{1}{4}$, as we show in boxes in Fig. 4.5. We then check the outcomes in E_2 for which e_1 occurs, and add their new weights, obtaining

$$\text{pr}[e_1 \mid e_2] = \tfrac{1}{4} + \tfrac{1}{4} = \tfrac{1}{2}.$$

In Example 4.2, we found that $\text{pr}[e_1] = \frac{1}{3}$, so e_1 and e_2 are not independent events. $\|$

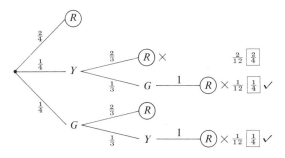

Fig. 4.5

EXERCISES

4.1 Find the probability that four flips of a fair coin produce four tails.

4.2 A biased coin is twice as likely to produce a head as a tail when flipped. Find the probability that three flips of the coin produce exactly two heads.

4.3 For the coin in Exercise 4.2, find the probability that three flips produce at least two heads, given that at least one head appears.

4.4 A bag contains two red gumdrops, one green gumdrop, and one yellow one. Gumdrops are drawn in succession at random without replacement until a red one is drawn, and then no more are drawn. Find the probability that both the green and yellow gumdrops are drawn.

4.5 Referring to Exercise 4.4, find the probability that both the green and yellow gumdrops are drawn, given that the yellow one is drawn.

4.6 A bag contains two red gumdrops, one green gumdrop, and one yellow one. Gumdrops are drawn in succession at random without replacement until the yellow one is drawn, and then no more are drawn. Find the probability that both red gumdrops are drawn.

4.7 For the experiment in Exercise 4.6, find the probability that both red gumdrops are drawn, given that the green gumdrop is drawn.

4.8 Two cards are drawn in succession without replacement from a deck of 52 playing cards. Find the probability that the cards are both of the same color.

4.9 A die is loaded so that when it is rolled the probability that a number occurs is proportional to that number. Thus a 6 is three times as likely to occur as a 2, etc. Find the probability that three rolls of the die produce exactly two even numbers.

4.10 For the die in Exercise 4.9, find the probability that the sum of the numbers which appear on three rolls of the die is an odd number.

4.11 In a simple genetic model, each individual has two genes which govern a certain physical trait (such as eye color). Each gene is either of type G or type g. An individual may then be classified with respect to this trait as either type GG (dominant), type Gg (hybrid), or type gg (recessive), according to whether it has two type G genes, one of each type, or two type g genes. An offspring inherits one of its genes from each parent, the gene which it inherits from a parent being a random choice of one of that parent's two genes. Find the probability that

a) an offspring of a hybrid parent and a recessive parent is recessive,
b) an offspring of a dominant parent and a recessive parent is hybrid,
c) an offspring of a dominant parent and a hybrid parent is recessive.

4.12 Referring to Exercise 4.11, find the probability that an offspring of two offspring of a dominant parent and a hybrid parent is hybrid.

5. INDEPENDENT TRIALS

Repeated flipping of a coin is an example of an independent-trials process. Each flip of the coin is a *trial*. For each trial, the outcomes H and T are the same and the weights of these outcomes are the same. Thus each trial is *independent* of all the other trials. Similarly, repeated rolling of a die is an independent-trials process. However, our gumdrop examples in the preceding section were not independent-trials processes, for the outcomes and also the weights varied from one stage of the tree to another.

An independent-trials process is naturally analyzed by a tree, and is completely determined by the outcomes and branch weights of the first stage of the tree, and by the number of trials performed. The first stage branching of the tree is simply repeated over and over for as many trials as are performed.

5.1 Trials with Two Outcomes

First we consider independent trials with just two outcomes. Since the outcomes must be complementary outcome events, we may as well denote them by s (*success*) and $\bar{s}$ (*failure*). On each trial, let $\mathrm{pr}[s] = p$, so that $\mathrm{pr}[\bar{s}] = 1 - p$.

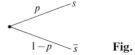

Fig. 5.1

It is easy to find the probability that n trials produce the outcome s exactly r times, that is, produce exactly r successes. Consider a tree for these n trials. The first stage is shown in Fig. 5.1. If we imagine that we draw the whole tree, we would then want to check the paths on which the outcome s occurs exactly r times. Such a path with exactly r outcomes s on it must have $n - r$ outcomes $\bar{s}$ also. The weight of any path is the product of its branch weights, so we see that every path we would check would have weight

$$p^r(1 - p)^{n-r},$$

where each factor p comes from a branch leading to an outcome s and each factor $1 - p$ comes from a branch leading to an outcome $\bar{s}$. There will be one such path for each way that r outcomes s and $n - r$ outcomes $\bar{s}$ can be distributed among n trials. Such a distribution is completely determined by selecting the r trials where the outcomes s occur from the n trials. Thus there would be $\binom{n}{r}$ paths checked. In summary, we would check $\binom{n}{r}$ paths, each of weight $p^r(1 - p)^{n-r}$, so the probability of obtaining exactly r successes in n trials is $\binom{n}{r}p^r(1 - p)^{n-r}$. We state this as a theorem, and give some examples.

> **Theorem 5.1** *For n independent trials with two outcomes s (success) and $\bar{s}$ (failure), where* $\mathrm{pr}[s] = p$ *for each trial, the probability that exactly r successes occur is* $\binom{n}{r}p^r(1 - p)^{n-r}$.

Example 5.1 A fair coin is flipped five times. Let us find the probability that exactly three heads are obtained.

For each trial, we let s be the outcome event

A head is obtained.

We have $\mathrm{pr}[s] = \frac{1}{2}$. By Theorem 5.1, our desired probability is

$$\binom{5}{3}\left(\frac{1}{2}\right)^3\left(1 - \frac{1}{2}\right)^{5-3} = \binom{5}{3}\left(\frac{1}{2}\right)^3\left(\frac{1}{2}\right)^2 = \frac{10}{32} = \frac{5}{16}. \;\|$$

Example 5.2 A biased coin is twice as likely to produce a head as a tail. Let us find the probability that five flips of the coin produce at least three heads.

For each trial, we let s be the outcome event

A head is obtained.

The formula in Theorem 5.1 is geared to finding the probability of obtaining *exactly r* successes in *n* trials. To find the probability of obtaining *at least* three successes in five trials, we find the probability that exactly three *or* exactly four *or* exactly five successes occur; we must add these individual probabilities. These probabilities are: For exactly three outcomes s,

$$\binom{5}{3}\left(\frac{2}{3}\right)^3\left(\frac{1}{3}\right)^2 = \frac{80}{3^5} \;;$$

for exactly four outcomes s,

$$\binom{5}{4}\left(\frac{2}{3}\right)^4\left(\frac{1}{3}\right)^1 = \frac{80}{3^5} \;;$$

and for exactly five successes,

$$\binom{5}{5}\left(\frac{2}{3}\right)^5\left(\frac{1}{3}\right)^0 = \frac{32}{3^5}.$$

The sum of these gives us $192/3^5$ or $64/81$ for our desired probability. $\|$

Example 5.3 Let a fair die be rolled four times, and let us find the probability that a 4 appears exactly twice.

This might seem like an independent-trials process with more than two outcomes, but we are only interested in whether or not a 4 appears on each trial, so we can take as outcomes for each trial

s: *The number 4 appears,*

$\bar{s}$: *A number different from 4 appears.*

Then $\mathrm{pr}[s] = \frac{1}{6}$, and $\mathrm{pr}[\bar{s}] = \frac{5}{6}$. By Theorem 5.1, the probability of obtaining the number 4 exactly twice in four rolls of the die is

$$\binom{4}{2}\left(\frac{1}{6}\right)^2\left(\frac{5}{6}\right)^2 = 6 \cdot \frac{25}{6^4} = \frac{25}{6^3} = \frac{25}{216}. \;\|$$

5.2 Trials with More than Two Outcomes

For an independent-trials process with more than two outcomes, we can find the probability that a certain distribution of the outcomes occurs in n trials by using the reasoning which led us to Theorem 5.1. That is:

STEP 1. Imagine the tree for the n trials to be drawn. (We suggest that you draw the first stage.)

STEP 2. Find the weight of each path which has the desired distribution of outcomes.

STEP 3. Find the number of paths which have the desired distribution of outcomes.

STEP 4. Compute the product of the numbers found in Steps 2 and 3.

We illustrate with an example.

Example 5.4 A fair die is rolled four times. Let us find the probability that exactly two even numbers and exactly one number 5 appear.

STEP 1. For the purpose of this problem, we can consider an independent-trials process with three outcomes. The first stage of the tree is shown in Fig. 5.2.

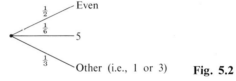

Fig. **5.2**

STEP 2. Each path of the whole tree with exactly two even outcomes and one outcome 5 also has exactly one outcome *other*, since there are four trials in all. The weight of such a path is

$$\left(\frac{1}{2}\right)^2 \left(\frac{1}{6}\right)^1 \left(\frac{1}{3}\right)^1 = \frac{1}{2^2 \cdot 6 \cdot 3}.$$

STEP 3. The number of paths with the desired distribution is the number of ways that from four trials you can select two where an even number is to appear, *and* then select one from the remaining two where a 5 is to appear. Thus there are

$$\binom{4}{2}\binom{2}{1} = 12$$

paths with the desired distribution of outcomes.

STEP 4. We find from Steps 2 and 3 that the desired probability is

$$12 \cdot \frac{1}{2^2 \cdot 6 \cdot 3} = \frac{1}{6}. \; \|$$

EXERCISES

5.1 Find the probability that six flips of a fair coin produce exactly four heads.

5.2 Find the probability that six flips of a fair coin produce at least four heads.

5.3 What is the most likely number of heads to appear in six flips of a fair coin?

5.4 Find the probability that four rolls of a fair die produce exactly one 3.

5.5 Find the probability that four rolls of a fair die produce no numbers 3.

5.6 Find the probability that four rolls of a fair die produce at most two numbers 3.

5.7 Mr. Smith has either bacon or sausage (but not both) for breakfast each morning. Given that he is as likely to have one as the other each day, find the probability that during the course of a week, he has bacon at least three days. [*Hint:* It is easier to find the probability that he has bacon at most two days.]

5.8 A fair die is rolled three times. Find the probability that exactly two numbers 4 and exactly one number 2 appear.

5.9 A fair die is rolled five times. Find the probability that more even than odd numbers appear.

5.10 A fair die is rolled four times. Find the probability that exactly one odd number and exactly two numbers 6 appear.

5.11 A fair die is rolled four times. Find the probability that exactly one odd number and exactly two numbers 5 appear.

5.12 A fair die is rolled four times. Find the probability that exactly three even numbers and exactly one 6 appear.

5.13 A fair die is rolled four times. Find the probability that exactly three even numbers appear, given that exactly one 6 appears. [*Hint:* Use the conditional probability formula here.]

5.14 On each Saturday night that he has a date, Bill dates either a redhead, a blonde, or a brunette. He has a date an average of five Saturdays out of every six, and averages equal numbers of blondes and brunettes, but only half as many redheads on the average.

a) Find the probability that Bill dates exactly one blonde and exactly two brunettes in a period of five Saturdays.

b) Find the probability that during a period of five consecutive weeks Bill has dates on precisely the first, third, and fifth Saturdays.

c) Find the probability that Bill dates more blondes than brunettes during a period of four Saturdays.

6. EXPECTED VALUE

6.1 An Illustration

We commence this section with an illustration concerning expected value.

Example 6.1 Suppose that Smith bets Jones that a certain event e will occur. Perhaps Smith bets his $3.00 against Jones' $2.00, and perhaps $pr[e] = \frac{2}{3}$. We shall compute the *expected value of this bet for Smith*.

The appropriate tree with outcomes e and $\bar{e}$ is shown in Fig. 6.1. Next to each outcome, we have indicated the value to Smith if this outcome occurs, for we are interested in how Smith fares with this bet. If $\bar{e}$ occurs, then Smith loses his \$3.00, and we have indicated this *loss* in Fig. 6.1 by placing "— \$3.00" next to $\bar{e}$.

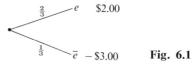

e \$2.00

$\bar{e}$ — \$3.00   **Fig. 6.1**

By the *expected value of this bet for Smith*, we mean the amount that Smith would expect to *average per bet* if he made this same bet over and over with Jones. Since $\mathrm{pr}[e] = \frac{2}{3}$, Smith can expect e to occur about two-thirds of the time, and each time e occurs, he wins \$2.00. Thus Smith expects the outcome e to contribute an average of $\frac{2}{3}(\$2.00)$ per bet to his pocket. Similarly, he expects to lose \$3.00 one-third of the time, so outcome $\bar{e}$ can be expected to take an average of $\frac{1}{3}(\$3.00)$ per bet from his pocket. In terms of *contributions* to his pocket, $\bar{e}$ contributes the negative amount $\frac{1}{3}(-\$3.00)$ per bet. Thus the total expected contribution, per bet, to Smith's pocket is the sum

$$\tfrac{2}{3}(\$2.00) + \tfrac{1}{3}(-\$3.00) = \frac{\$1.00}{3}.$$

This value, $\frac{1}{3}$ of a dollar, is the *expected value of the bet for Smith*. ‖

In the preceding example, note that Smith cannot win his expected value of $\frac{1}{3}$ of a dollar in his single bet; he either wins \$2.00 or loses \$3.00. Obviously this bet is favorable to Smith, and unfavorable to Jones. Of course, the expected value of the bet for Jones is $-\frac{1}{3}$ of a dollar, indicating that Jones must pay Smith an average of $\frac{1}{3}$ of a dollar per bet. A bet is **fair** if and only if the expected value for each person who bets is 0.

6.2 Random Variables

An essential feature of the tree in Fig. 6.1 is the assignment of a numerical value to each outcome event. The value 2 was assigned to e and the value -3 was assigned to $\bar{e}$. The assignment of a numerical value to each outcome event in a sample space S is of course a *function mapping S into* **R**; we introduced this concept in Chapter 1. These functions have a special name in the theory of probability.

Definition. Let S be a sample space with a probability measure. A *random variable on S* is a function $f \colon S \to \mathbf{R}$.

Thus if f is the random variable indicated by the assignment of values in the tree shown in Fig. 6.1, where the sample space is $\{e, \bar{e}\}$, we have $f(e) = 2$, while $f(\bar{e}) = -3$.

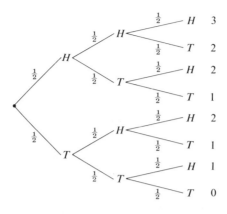

Fig. 6.2

Example 6.2 There are many cases other than betting situations in which random variables arise naturally. For example, when a fair coin is flipped three times, you may be interested in how many heads appear. Let f be the random variable which assigns, to each path outcome of the tree in Fig. 6.2, the number of heads which appear on the path. In Fig. 6.2 the value assigned to each path is shown next to its terminal point. For example, for the second path HHT, we have $f(HHT) = 2$. ‖

6.3 The Mean of a Random Variable

The *mean of a random variable* is the same thing as the expected value of the numerical quantities which the random variable assigns to the elements in the sample space. Thus for the bet in Example 6.1, the mean of the random variable f, where $f(e) = 2$ and $f(\bar{e}) = -3$, is $\frac{1}{3}$, which is the expected value for the bet. Let us give a precise definition of this concept. The definition is motivated by our computation of the expected value of the bet for Smith in Example 6.1.

> **Definition.** Let $S = \{e_1, \ldots, e_n\}$ be a sample space with a probability measure and let $f: S \to \mathbf{R}$ be a random variable. The **mean μ of f** is given by
>
> $$\mu = \mathrm{pr}[e_1] \cdot f(e_1) + \cdots + \mathrm{pr}[e_n] \cdot f(e_n).$$

If a random variable f assigns values $a_1, \ldots, a_n$ to elements $e_1, \ldots, e_n$ in a sample space for an experiment, then the **expected value of the a_i** is simply the mean of f. The phrases, *the mean value of* and *the mean number of*, are often used as synonyms for *the expected value of* and *the expected number of*, without reference to the specific random variable under consideration, if the random variable is clear from the context. Thus if a coin is flipped three times, we may speak of "the mean number of heads which appear," or, equivalently, "the expected number of heads which appear." In this case, the random variable is understood to be the one discussed in Example 6.2.

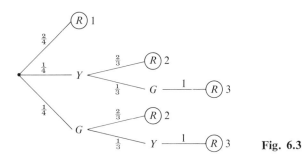

Fig. 6.3

We conclude with two more examples.

Example 6.3 A bag contains two red gumdrops, one yellow gumdrop, and one green one. Gumdrops are drawn in succession at random without replacement until a red gumdrop is obtained, and then no more are drawn. Let us find the expected number of draws required to complete this experiment.

A tree for the experiment is shown in Fig. 6.3, and next to each terminal point of each tree path we place the number of draws required to travel that path, for we are obviously interested in the mean of the random variable which assigns, to each path outcome, the number of draws which is required. By definition, we find that

$$\mu = (\tfrac{1}{2}) \cdot 1 + (\tfrac{1}{4} \cdot \tfrac{2}{3}) \cdot 2 + (\tfrac{1}{4} \cdot \tfrac{1}{3} \cdot 1) \cdot 3 + (\tfrac{1}{4} \cdot \tfrac{2}{3}) \cdot 2 + (\tfrac{1}{4} \cdot \tfrac{1}{3} \cdot 1) \cdot 3 = \tfrac{5}{3}.$$

Of course, it is impossible to make $\tfrac{5}{3}$ draws; you either draw one, two, or three times. However, if the experiment were repeated over and over, you would expect to average $\tfrac{5}{3}$ draws per experiment. For example, if 300 such experiments were performed using 300 bags, each containing two red gumdrops, one yellow gumdrop, and one green one, you would expect to make about $\tfrac{5}{3} \cdot 300 = 500$ draws in all. ‖

Example 6.4 Let us find the expected number of heads which occur in three flips of a fair coin.

Fig. 6.4

If we view the expected number as the average number of heads we would expect to appear per experiment (three flips) for many repetitions of the experiment, it is obvious that the expected number for three flips of the coin will be three times the expected number for a single flip. The tree for a single flip is given in Fig. 6.4, and we see that for one flip, the expected number of heads is

$$\tfrac{1}{2} \cdot 1 + \tfrac{1}{2} \cdot 0 = \tfrac{1}{2}.$$

Therefore, for three flips, we have an expected number of $\frac{3}{2}$ heads. Note that it was not necessary to apply our definition to the large tree shown in Fig. 6.2. ‖

The preceding example illustrates a useful property of mean values, namely:

For an independent-trials process, the mean value of a random variable for n trials is n times the corresponding mean value for a single trial.

EXERCISES

6.1 Jones has $1.80 to bet with Smith that a head will appear when he flips a coin. If Jones can flip a head two times out of three, how much money should Smith put up if the bet is to be fair?

6.2 Find the expected number of heads which appear in seven flips of a fair coin.

6.3 Find the expected number of heads which appear in seven flips of a coin which is biased so that a head appears on each flip with probability $\frac{1}{3}$.

6.4 Find the mean value of the sum of the numbers which appear on three rolls of a fair die. [*Hint:* Use the principle for an independent-trials process which was stated at the end of this section.]

6.5 A bag contains two red gumdrops, one yellow gumdrop, and one green one. Gumdrops are drawn successively at random without replacement until the green one is drawn, and then no more are drawn. Find the expected number of draws required.

6.6 For the experiment described in Exercise 6.5, find the expected number of draws required, given that both red gumdrops are drawn.

6.7 A die is loaded in such a way that each odd number is as likely to appear as any other odd number, but each even number is twice as likely to appear as an odd number.

a) Find the expected value of the number which appears in a single roll of the die.
b) Find the mean value of the sum of the numbers which appear in five rolls of the die.

6.8 Smith and Jones are cutting cards. If a red card appears, Smith pays Jones $2.00, while if a spade appears, Jones pays Smith $4.00. Is this a fair bet? If not, to whom is it favorable?

6.9 A bag contains two red gumdrops, one green gumdrop, and one yellow gumdrop. Gumdrops are drawn successively at random without replacement until the green gumdrop is obtained, and then no more are drawn.

a) Find the expected number of red gumdrops drawn.
b) Find the expected number of yellow gumdrops drawn.
c) Find the expected number of green gumdrops drawn.
d) How is the expected number of draws required to complete the experiment related to the answers to parts (a), (b), and (c)?

6.10 A man has two quarters, three dimes, and two nickels in his pocket. He draws two coins from his pocket in succession at random without replacement.

a) What is the expected value of the coin he obtains on the first draw?
b) What is the expected value of the coin he obtains on the second draw?
c) What is the expected total value of the two coins?
d) Devise a method for answering (a), (b), and (c) without drawing a tree.

6.11 For the experiment of Exercise 6.10, find the expected value of the first draw, given that the second draw produced a dime.

6.12 On each Saturday night that he has a date, Bill dates either a redhead, a blonde, or a brunette. He has a date an average of five Saturdays out of every six, and he averages equal numbers of blondes and brunettes, but only half as many redheads on the average. He figures that the cost of a date with a blonde is $8.00 (they always want more fun), the cost with a redhead is $6.00, and the cost with a brunette is $5.00. How much can he expect to spend on Saturday night dates during the course of a year (52 Saturdays)?

7. RANDOM WALKS

7.1 One-Dimensional Random Walks

Consider a man walking on a number line, as sketched in Fig. 7.1. Let us assume that his steps are always one unit long, and that he starts his walk at some integral initial position (perhaps 10 or 57) on the number line. As he takes each step, he has a choice of whether he should take the step in the positive direction (to the right) or in the negative direction (to the left) on the number line. For example, if he begins at position 10, then after one step he will be at either position 11 or position 9. Suppose that before he takes each step he flips a coin which comes up heads with probability p (and therefore comes up tails with probability $1 - p$). If the coin comes up heads, the man takes his step in the positive direction, while if the flip produces a tail, he take a step in the negative direction. This is a model for a ***one-dimensional random walk with probability*** p. If $p = \frac{1}{2}$, so that the man is as likely to take each step in one direction as in the other, then the random walk is ***symmetric***.

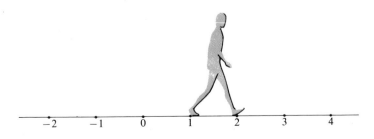

Fig. 7.1

Example 7.1 Suppose a man executes a random walk starting from position 7, and suppose that at each step he goes in the positive direction with probability $\frac{2}{3}$, that is, $p = \frac{2}{3}$. Let us find his expected position after 12 steps.

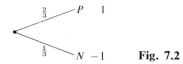

Fig. 7.2

This random walk is essentially an independent-trials process with two outcomes, namely P (a step in the *positive* direction) and N (a step in the *negative* direction) for each of 12 trials. The first stage of the appropriate tree is shown in Fig. 7.2. Let us take the random variable which assigns $+1$ to P (signifying one step to the right) and -1 to N (signifying a negative step to the right, i.e., one step in the negative direction). The expected number of (positive and negative) steps to the right for this first trial is

$$\tfrac{2}{3} \cdot (1) + \tfrac{1}{3}(-1) = \tfrac{1}{3}.$$

Thus in 12 steps the man expects to have taken a net total of

$$12 \cdot \tfrac{1}{3} = 4$$

steps to the right. Since he started at position 7, his expected position after 12 steps is 11. ‖

Random walks are important probabilistic models for studying diffusion and motion in particle physics. The physicist assumes that his particles are in motion, and that collisions between particles impart to them a motion which is similar to that of a random walk. For example, the particles might be gas molecules which are colliding with each other. Probability is becoming increasingly important in the study of the physical world. Some physicists feel that the most important models for studying the universe may turn out to be probabilistic ones.

Physical considerations motivate the introduction of a *random walk with barriers*. For example, a physicist may be studying diffusion or molecular motion of a gas in a container. The container forms a barrier for the molecules of gas. Barriers are of essentially three types.

1. A barrier may be such that if a particle hits the barrier, it passes through (or is trapped) and cannot return. Such a barrier is **absorbing**, and the random walk of a particle terminates if the particle hits an absorbing barrier.

2. A barrier may be such that if a particle hits the barrier, it bounces back away from the barrier to its previous position. Such a barrier is **reflecting**, and a random walk of a particle never terminates if all barriers are reflecting.

3. A barrier may be such that if the particle hits the barrier hard enough, it passes through and is absorbed, but if it hits the barrier at a lower speed, it is reflected. Such a barrier is *elastic*.

A *restricted random walk* is one with barriers; an ***unrestricted random walk*** has no barriers.

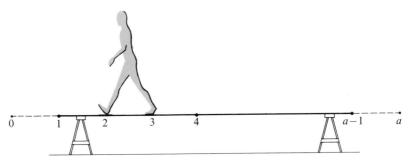

Fig. 7.3

As a model for a one-dimensional random walk with absorbing barriers, let us consider a man executing a random walk on a plank resting on two sawhorses, as shown by the heavy line in Fig. 7.3. We imagine that the plank is numbered from 1 to $a - 1$ for some $a \in \mathbf{Z}^+$, as indicated in the figure. This model corresponds to a random walk on the number line with absorbing barriers at 0 and a. If the man is at position 1 and tries to take a step in the negative direction to 0, he falls off and is absorbed, that is, his walk terminates. Similarly, he is absorbed at a if he tries to take a step from $a - 1$ to a.

Example 7.2 Suppose a man starts at position 2 and executes a symmetric one-dimensional random walk on the number line with absorbing barriers at 0 and 3. Let us find the probability that he has not been absorbed after three steps.

The tree for three steps of this walk is shown in Fig. 7.4. If the man reaches 0 or 3, he is absorbed (falls off) and the walk terminates. There is only one path of the tree he could take and not be absorbed after three steps; the probability that he follows this path is $\frac{1}{8}$. ‖

Example 7.3 Suppose a particle with initial position 2 executes a symmetric one-dimensional random walk on the number line with elastic barriers at $\frac{1}{2}$

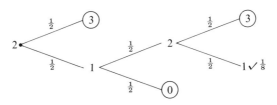

Fig. 7.4

and $\frac{5}{2}$. The barriers are such that if the particle hits a barrier, it is as likely to be absorbed as to be reflected. For example, if the particle is at 1, it may take a step in the positive direction to 2 with probability $\frac{1}{2}$, or it may try to take a step in the negative direction and either pass the barrier and be absorbed at 0 with probability $\frac{1}{4}$ or be reflected back to 1 by the barrier, again with probability $\frac{1}{4}$. Let us find the probability that the particle is not absorbed after two steps.

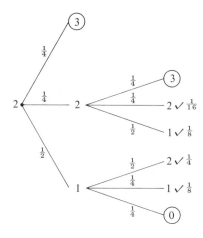

Fig. 7.5

The tree for two steps of the walk is shown in Fig. 7.5. Again we have checked the paths where the particle is not absorbed, and we find that the probability that it is not absorbed is

$$\frac{1}{16} + \frac{1}{8} + \frac{1}{4} + \frac{1}{8} = \frac{1 + 2 + 4 + 2}{16} = \frac{9}{16}. \;\|$$

The two preceding examples suggest the following theorem, which we shall prove for you.

Theorem 7.1 *If a particle executes a one-dimensional random walk bounded by absorbing or elastic barriers, then the probability that the particle is absorbed approaches 1 as the number of steps becomes large. Thus if the walk is allowed to continue indefinitely, the particle is absorbed with probability 1.*

Proof. By assumption, the walk is bounded by barriers at which the particle can be absorbed. Suppose that it is possible for the particle to reach a barrier from *any* initial position in n steps. Then there is some number $r > 0$ such that the probability that the particle is absorbed after n steps from any position is at least r. The probability that the particle is *not* absorbed after n steps is therefore at most $(1 - r) < 1$. Elementary arguments with a tree (see Fig. 7.6) show that the probability that the particle is not absorbed after

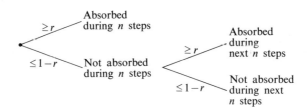

Fig. 7.6

$2n$ steps is $\leq (1 - r)^2$, and, in general, the probability that the particle is not absorbed after mn steps is $\leq (1 - r)^m$. As m gets large, $(1 - r)^m$ approaches 0 since $(1 - r) < 1$. Thus as the number of steps increases, the probability that the particle is *not* absorbed approaches 0, so the probability that it is absorbed must approach 1. ∎

7.2 Two- and Three-Dimensional Random Walks

Imagine a particle in a plane. Suppose that the particle may, with certain probabilities, take a unit step in one of four directions:

forward,   backward,   right,   left.

We view this as one step of a **two-dimensional random walk**. We may think of the plane as $\mathbf{R}^2$, and if the particle is at $(1, 2)$, then it may move to $(1, 3)$, $(1, 1)$, $(2, 2)$, or $(0, 2)$ in one step (see Fig. 7.7). The walk is **symmetric** if the probability is $\frac{1}{4}$ for a step in each of these four directions, at each stage of the walk. Similarly, a particle in space (which we may view as $\mathbf{R}^3$) executes a **three-dimensional random walk** by taking successive unit steps, with certain probabilities, in the six directions:

up,   down,   forward,   backward,   right,   left.

If the particle starts at $(3, 1, -2)$, then after one step it will be at one of the

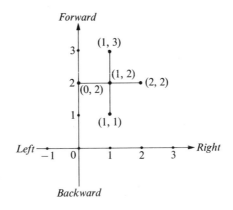

Fig. 7.7

points $(4, 1, -2)$, $(2, 1, -2)$, $(3, 2, -2)$, $(3, 0, -2)$, $(3, 1, -1)$, or $(3, 1, -3)$. For a *symmetric walk*, the probability must be $\frac{1}{6}$ for a step in each of the six directions at each stage of the walk. Barriers may be introduced as in one-dimensional walks, and again an *unrestricted walk* is one without barriers.

We introduced these higher-dimensional walks simply to state a fascinating result due to G. Polya. An interesting question for random walks is whether or not a particle ever returns to its initial position during the walk. It would seem reasonable that a particle in a symmetric random walk (of either one, two, or three dimensions) should return to its initial position, since the expected position after any number of steps is easily seen to be the initial position (see Exercise 7.4). However, Polya proved the following fascinating theorem which separates one- and two-dimensional symmetric walks from three-dimensional symmetric walks in this respect.

Theorem 7.2 (Polya). *For one- and two-dimensional unrestricted symmetric random walks, a particle returns to its initial position with probability 1. However, for a three-dimensional unrestricted symmetric random walk, the probability of the particle's returning to its initial position is only about 0.35.*

It is easy to argue from Polya's theorem that for a one- or two-dimensional unrestricted symmetric random walk, a particle returns to its initial position infinitely often with probability 1 (see Exercise 7.9).

EXERCISES

7.1 A drunken man is standing on a dock and is facing in the direction of the short dimension of the dock, i.e., parallel to the shoreline. He will fall off the dock into the water on his second step forward or on his second step backward from his initial position. If he executes a random walk in which he is as likely to take a step forward as backward, what is the probability that he will still be on the dock after three steps?

7.2 A particle at position 2 on the number line starts a random walk, moving in the positive direction at each stage with probability $p = \frac{2}{3}$. The walk has absorbing barriers at 0 and 4. Find the probability that the particle is absorbed on or before the fourth step.

7.3 A particle at position 2 on a number line starts a random walk, moving in the positive direction at each stage with probability $p = \frac{2}{3}$. The line has reflecting barriers at $\frac{1}{2}$ and $\frac{7}{2}$ which bounce the particle back to 1 and 3, respectively. Find the expected position of the particle after three steps.

7.4 Show that for an unrestricted symmetric random walk in one, two, or three dimensions, the expected position after n steps is the initial position.

7.5 A particle at position 3 on a number line starts a symmetric random walk on the line. There are elastic barriers at $\frac{1}{2}$ and $\frac{9}{2}$ which are twice as likely to absorb a particle as to reflect it back (to 1 and 4, respectively). Find the probability that the particle is absorbed on or before its third step.

7.6 A particle in the Euclidean plane $\mathbf{R}^2$ starts at $(1, 0)$ and executes a symmetric random walk. There is an absorbing barrier at $\{(x, y) \mid y = 1\}$ and another at $\{(x, y) \mid y = -1\}$. Find the probability that the particle has not yet been absorbed after three steps.

7.7 A particle in the Euclidean plane $\mathbf{R}^2$ starts at $(1, 0)$ and executes a symmetric random walk. There is an elastic barrier at $\{(x, y) \mid y = \frac{1}{2}\}$ which is three times as likely to reflect a particle as to absorb it, and another elastic barrier located at $\{(x, y) \mid y = -\frac{1}{2}\}$ which is twice as likely to absorb a particle as to reflect it. Find the probability that the particle has not yet been absorbed after two steps.

7.8 Is Theorem 7.1 valid for two- and three-dimensional random walks also? [*Hint:* Try an analogous argument.]

7.9 Argue from Theorem 7.2 that for an unrestricted symmetric random walk in one or two dimensions, a particle will return to its initial position infinitely often with probability 1.

7.10 Argue that for an unrestricted symmetric random walk in one or two dimensions, every position is reached sometime with probability 1. [*Hint:* Use the result of Exercise 7.9, and apply the technique of Theorem 7.1 to show that for any position the probability that the position is not reached approaches 0 as the walk continues indefinitely.]

7.11 Suppose two particles start at the same time at positions 2 and 7 on a number line and execute simultaneous unrestricted random walks on the number line, taking their unit steps at the same time. Show that the particles never occupy the same position simultaneously. [*Hint:* Make an even-odd argument.]

7.12 For the situation in Exercise 7.11, suppose that the initial positions are 2 and 8 and suppose that the walk of each particle is symmetric. Argue that the particles occupy some position simultaneously with probability 1. [*Hint:* Form a new random walk with the position number given by the distance between the particles, and use Exercise 7.10.]

7.13 Two gamblers, Smith and Jones, are engaged in a game of chance. At each play of the game, Smith wins a dollar from Jones with probability p, and loses a dollar to Jones with probability $1 - p$. The game continues until either Smith or Jones is ruined, i.e., has no money left. (This probabilistic model is known as *gambler's ruin*.) Suppose that Smith has an initial capital of A dollars and Jones has an initial capital of B dollars.

a) If Jones is ruined, how much money will Smith then have?

b) If, after a while, Smith has a total of X dollars, where $0 < X < A + B$, how much might Smith have after the next play of the game, and with what probabilities?

c) Regarding Smith's capital after each play of the game as his numerical position, show that this gambler's ruin can be regarded as a random walk which starts at position A on a number line, has probability p of taking a step in the positive direction at each stage, and has absorbing barriers at 0 and $A + B$.

7.14 With reference to Exercise 7.13, if Smith has $2.00 and Jones has $1.00 at the start of the game, and if the game is a fair one, what is the probability that Jones will be ruined within three plays of the game?

You will find an illuminating and elementary discussion of gambler's ruin in Kemeny, Snell, and Thompson [3, page 209]. It is not necessary to know what a Markov chain is to understand this reference.

8. THE CENTRAL LIMIT THEOREM

8.1 Introduction

As our work in Section 6 showed, the expected number of heads resulting from n flips of a fair coin is $n/2$. You should not think that this means that in n flips you will be very likely to get *exactly* $n/2$ heads. Of course, if n is odd, the probability of getting exactly $n/2$ heads is 0. Even if n is even, the probability of *exactly* $n/2$ heads is close to 0 if n is large. For example, while the probability of obtaining exactly one head in two flips is 0.5, the probability of getting exactly five heads in ten flips is down to about 0.25. If you think about it for a moment, it is clear that the probability of getting *exactly* 500,000 heads in a million flips of a fair coin is very small indeed. What is true is that in n flips of a fair coin, the probability is close to 1 that the number of heads won't deviate too far from $n/2$. Let us discuss what we mean by this statement.

In ten flips of a fair coin, the probability that either three, four, five, six, or seven heads appear is about 0.89, as you can verify using the techniques of Section 5. We can rephrase this by saying that with probability about 0.89, in ten flips of a fair coin, the number of heads which appear deviates from the mean number 5 by at most 2. Again, a little thought will convince you that, for a million flips of a fair coin, the probability is nowhere near as large as 0.89 that the number of heads which appear deviates from the mean number 500,000 by at most 2; the probability is very close to 0. The probabilities of obtaining exactly 499,998, exactly 499,999, exactly 500,000, exactly 500,001, and exactly 500,002 heads are all very close to zero, and their sum is still very close to zero.

We would like to have some sort of a *unit of deviation* such that the probability that, for ten flips of a fair coin, the number of heads differs from the mean number 5 by r of these units is about the same as the probability that, for a million flips of the coin, the number of heads differs from the mean number 500,000 by r of these same units. Our arguments above show that a unit of *one head* is not going to work, for with $r = 2$, we had very different probabilities for ten flips and for a million flips. The type of unit of deviation which we are hunting for is going to have to encompass more heads for a million flips than it does for ten flips.

Let us make one more "natural" attempt. The size of the number n of flips clearly plays a role. Perhaps an appropriate unit of deviation would be a fraction of n, say *one percent of n*, that is, $n/100$. For example, if $n = 10$, then ten of these units would amount to

$$10 \cdot \tfrac{10}{100} = 1 \text{ head,}$$

and for $n = 1,000,000$, ten of these same units would amount to

$$10 \cdot \frac{1,000,000}{100} = 100,000 \text{ heads.}$$

Perhaps the probability of obtaining from four to six heads in ten flips is about the same as the probability of obtaining from 400,000 to 600,000 heads in a million flips of the coin. Both of these ranges represent a deviation from the mean of at most ten of our *one percent of n* units. It is easy for us to compute that the probability of obtaining four, five, or six heads in ten flips of a fair coin is about 0.66. While we do not compute it for you now, the probability of obtaining between 400,000 and 600,000 heads in a million flips is greater than 0.99999999. Thus this attempt does not work either; while we need a unit of deviation that will encompass an increasing number of heads as the number n of flips increases, the unit *one percent of n* gets large too fast.

8.2 The Central Limit Theorem

At this point, we shall stop our exploratory discussion and tell you the answer to this question. We hope that the problem is clear to you. Let us state it in terms of any independent-trials process with two outcomes.

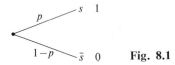

Fig. 8.1

Consider an independent-trials process with the two outcomes s (*success*) and $\bar{s}$ (*failure*), where for each trial, $\text{pr}[s] = p$. The mean number of outcomes s in a single trial is $1 \cdot p + 0(1 - p) = p$. (See the tree in Fig. 8.1.) Thus the mean number μ of successes in n trials is np.

PROBLEM. Is there a unit of deviation which varies with n such that, in n trials, the probability that the number of successes differs from the mean number $\mu = np$ by at most r of these units remains about the same as n varies?

The answer to our problem is *yes*, provided that n isn't allowed to get too small, and $\sqrt{np(1 - p)}$ *is such a unit of deviation*. For a given independent-trials process, where p remains constant, the quantity $\sqrt{np(1-p)}$ varies like $\sqrt{n}$. Since we saw above that the unit *one percent of n*, that is, $n/100$, increases too rapidly as n increases, your next natural attempt might well have been $\sqrt{n}$. It is customary to let $q = 1 - p$. We shall let $d = \sqrt{npq}$.

Definition. The ***unit d of standard deviation*** for n independent trials with two outcomes in each trial, occurring with probabilities p and $q = 1 - p$, is $\sqrt{npq}$.

The fact that the unit of standard deviation $\sqrt{npq}$ does solve our problem is a special case of one of the most important theorems of probability theory, the *Central Limit Theorem*. We state the special case we shall use.

Theorem 8.1 (Central Limit Theorem, special case). *Let s (success) and $\bar{s}$ (failure) be the outcomes in an independent-trials process with two outcomes, and let* $\mathrm{pr}[s] = p$ *for each trial. The probability that, in n trials, the number of successes differs from the mean number $\mu = np$ by not more than r of the units $d = \sqrt{npq}$ of standard deviation is close to a number $N(r)$ for all n sufficiently large. This number $N(r)$ depends only on r, and is independent of the value of p.*

It can also be shown that, for sufficiently large n, the probability that the number of successes falls in the interval $[\mu, \mu + rd]$ is the same as the probability that the number of successes falls in the interval $[\mu - rd, \mu]$. That is, the outcomes s are distributed evenly about the mean $\mu = np$. Let us denote by "PR$[r]$" the probability that the number of successes falls in the interval $[\mu, \mu + rd]$. The number $N(r)$ of Theorem 8.1 is the approximate probability that, for large n, the number of successes falls in the interval $[\mu - rd, \mu + rd]$, and is thus $2 \cdot$ PR$[r]$. In Fig. 8.2 we indicate on a number line the mean μ together with three units of standard deviation to the left and to the right of μ. We have also indicated the probability, for large n, of falling within certain intervals in terms of our "PR$[r]$" notation. A table giving the values of PR$[r]$ as r varies by tenths is given in Fig. 8.3.

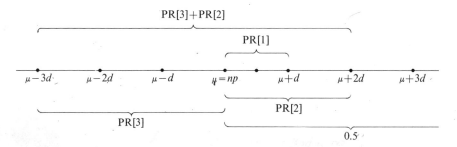

Fig. 8.2

r	PR$[r]$	r	PR$[r]$	r	PR$[r]$	r	PR$[r]$
0.0	0.000	1.1	0.364	2.1	0.482	3.1	0.4990
0.1	0.040	1.2	0.385	2.2	0.486	3.2	0.4993
0.2	0.079	1.3	0.403	2.3	0.489	3.3	0.4995
0.3	0.118	1.4	0.419	2.4	0.492	3.4	0.4997
0.4	0.155	1.5	0.433	2.5	0.494	3.5	0.4998
0.5	0.191	1.6	0.445	2.6	0.495	3.6	0.4998
0.6	0.226	1.7	0.455	2.7	0.497	3.7	0.4999
0.7	0.258	1.8	0.464	2.8	0.497	3.8	0.49993
0.8	0.288	1.9	0.471	2.9	0.498	3.9	0.49995
0.9	0.316	2.0	0.477	3.0	0.4987	4.0	0.49997
1.0	0.341					5.0	0.49999997

Fig. 8.3

For the sake of those who know a little integral calculus, we mention that

$$PR[r] = \frac{1}{\sqrt{2\pi}} \int_0^r e^{-x^2/2} \, dx,$$

where e is *not* an event, but rather a particular real number whose value is about 2.71828. In general, the probability that, for large n, the number of successes falls in the interval $[\mu - rd, \mu + td]$ is about

$$\frac{1}{\sqrt{2\pi}} \int_{-r}^t e^{-x^2/2} \, dx.$$

The graph of the function f such that $f(x) = (1/\sqrt{2\pi})e^{-x^2/2}$ is shown in Fig. 8.4. Thus PR$[r]$ is equal to the area of the shaded region shown in Fig. 8.4. Since $f(0) = 1/\sqrt{2\pi}$, we have made the scale on the y-axis greater than that on the x-axis in Fig. 8.4, so that the shaded region is easier to see. This "bell-shaped curve" plays an important roll in probability.

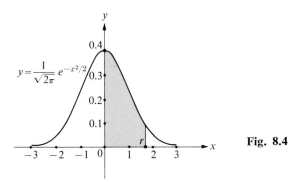

Fig. 8.4

The Central Limit Theorem states a result for "all n sufficiently large." It can be shown that so long as p is not too close to 0 or 1 and you are working in an interval near the mean μ, the table in Fig. 8.3 gives reasonable accuracy for n as low as 20. In general, $\mu = np$ should not be much less than 10 if the Central Limit Theorem is to be applied.

We conclude with some examples.

Example 8.1 A fair coin is flipped 100 times. Let us estimate the probability of obtaining between 45 and 55 heads.

We have $n = 100$ and $p = \frac{1}{2}$, so $\mu = np = 50$ and $d = \sqrt{npq} = \sqrt{100(\frac{1}{2})(\frac{1}{2})} = 10 \cdot \frac{1}{2} = 5$. Thus we want the probability that the number of heads falls within one unit of standard deviation of the mean, as indicated in Fig. 8.5. From our table in Fig. 8.3, we can attach the weights to the intervals as shown in Fig. 8.5. Thus our desired probability is about $0.341 + 0.341 = 0.682$. ‖

Example 8.2 For a million flips of a fair coin, we have $n = 1{,}000{,}000$, $p = \frac{1}{2}$, $np = 500{,}000$, and $d = \sqrt{npq} = \sqrt{1{,}000{,}000 \cdot \frac{1}{2} \cdot \frac{1}{2}} = 1000 \cdot \frac{1}{2} = 500$. Thus the probability of obtaining between 499,500 and 500,500 heads in a million flips is about 0.682, just as in Example 8.1, for this again corresponds to falling within one unit of standard deviation of the mean. ‖

Example 8.3 We stated in our introductory discussion for this section that the probability of obtaining between 400,000 and 600,000 heads in a million flips of a fair coin is very close to 1. By Example 8.2, the unit of standard deviation is 500. Thus the range of values from 400,000 to 600,000 is the range within 200 units of standard deviation of the mean! As the table in Fig. 8.3 indicates, this probability must be extremely close to 1. ‖

Example 8.4 Suppose the population of a very large city is $\frac{1}{3}$ republicans and $\frac{2}{3}$ democrats. For a random selection of 5000 people from the city, let us find the probability of obtaining fewer than 3350 democrats.

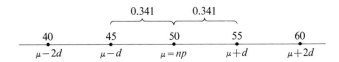

Fig. 8.5

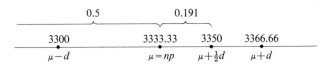

Fig. 8.6

For the selection of a single individual from the city, let *s* be the event

A democrat is chosen.

Then pr[*s*] = ⅔. Since our city is very large, we can view this as an independent-trials process with *n* = 5000 trials and *p* = pr[*s*] = ⅔. Then $\mu = np = 3333.33$, approximately, and $d = \sqrt{npq} = \sqrt{5000 \cdot \frac{2}{3} \cdot \frac{1}{3}} = \frac{100}{3} = 33.33$, approximately. The number 3350 is about 16.66 more than the mean, or 0.5 units of standard deviation more than the mean. We have indicated this data on a number line in Fig. 8.6, and have also indicated the weights of the intervals in which we are interested, as determined from the table in Fig. 8.3. Thus the desired probability is about 0.191 + 0.5 = 0.691. ∥

EXERCISES

8.1 For an independent-trials process with two outcomes *s* and *s̄*, use the table in Fig. 8.3 to find, for a large number of trials, the probability that the number of successes *s* falls within each of the indicated intervals of units of standard deviation from the mean.

a) [0, 0.3]   b) [−0.3, 0.3]   c) [−1, 0.5]   d) [−1, −0.5]
e)]−3, 2[f) >1.7   g) <−2.1   h) >−1.4

8.2 For the situation described in Exercise 8.1, explain why the answer for the interval]−3, 2[of units of standard deviation from the mean would be the same as the number for the interval [−3, 2].

8.3 Estimate the probability that 100 flips of a fair coin produce between 45 and 60 heads.

8.4 Estimate the probability that 200 rolls of a fair die produce at least 70 numbers greater than 4.

8.5 Smith and Jones are cutting cards at random. If the cut produces a red card, Smith pays Jones $1.00; if a black card appears, Jones pays Smith $1.00. Find the approximate probability that, after 400 plays, one of the players has lost more than $25.00.

8.6 A die is loaded in such a way that any odd number is as likely to come up as any other odd number, but any even number is twice as likely to come up as any odd number. Estimate the probability that 5000 rolls of the die produce at least 540 threes.

8.7 A particle executes a one-dimensional symmetric unrestricted random walk. Estimate the probability that, after 900 steps, the particle is within 20 units distance of its initial position. (All steps are one unit in length.)

8.8 An airplane drops 300 propaganda leaflets at random over an area of four acres.

a) Estimate the probability that each acre receives exactly 75 leaflets.
b) Estimate the probability that one particular acre, say the "north acre," receives less than 70 leaflets.

8.9 In an attempt to determine what proportion of a large population has type A blood, a sample of 1200 people is chosen at random and the blood type of each person is checked. If actually 25 percent of the population has type A blood, what is the approximate probability that more than 330 of the people checked have type A blood?

8.10 Prior to an election for mayor in a large city, half the voters favor candidate Smith for mayor and half favor candidate Jones. If opinion pollsters interview a random sample of 1600 voters, what is the probability that they will decide that more than 52 percent of the voters favor Jones?

8.11 A certain lazy professor decides to give a 225-question true-false test as final examination in his large freshman course. (The test can be graded by machine.) If a passing grade on the exam is 60 percent correct, what is the approximate probability that a student who simply guesses on every question will pass the examination? (The student is not penalized for wrong answers.)

8.12 Referring to Exercise 8.11, find the probability that the student passes the test if he knows the answers to 81 questions and guesses on the others.

9. CONFIDENCE INTERVALS

In this section we shall continue to study an independent-trials process with two outcomes, s (*success*) and $\bar{s}$ (*failure*). As before, we let $p = \text{pr}[s]$ be the probability of success on a single trial. The mean number of successes in n trials is $\mu = np$.

Frequently the mean μ and the probability p for an independent-trials process are not known, and the purpose of undertaking an experiment of n trials is to estimate them. We would expect n trials to produce about $\mu = np$ successes, and if the number of successes actually obtained is x, we regard x as the **empirical mean for the n trials**. Since $p = \mu/n$, we take as the **empirical probability for success on each trial**

$$p' = \frac{x}{n}.$$

Example 9.1 Let a coin be flipped 100 times. It is not known whether or not the coin is fair. For each flip, let s be the event

A head appears.

Suppose that $x = 60$ heads were obtained in the 100 flips. Then 60 is our empirical mean from this experiment, and the empirical probability p' is

$$\frac{x}{n} = \frac{60}{100} = 0.6. \;\|$$

The purpose of this section is to discuss the confidence that you can have in empirical values of the mean and probability which are determined by experimentation.

9.1 Confidence Intervals for the Mean

The Central Limit Theorem tells us that, for n sufficiently large, the probability that x differs from μ by at most r units d of standard deviation is approximately $N(r)$, where $N(r) = 2 \cdot \text{PR}[r]$, the value of $\text{PR}[r]$ being given in the table in Fig. 8.3. We express this symbolically by

$$\text{pr}[\mu - rd \leq x \leq \mu + rd] \approx N(r) \qquad (1)$$

for n sufficiently large, where "$\approx$" means "is approximately equal to." The inequality

$$\mu - rd \leq x \leq \mu + rd \qquad (2)$$

can be rewritten so that μ appears in the middle rather than x. From (2), by adding $-\mu - x$ to each term, we obtain

$$-x - rd \leq -\mu \leq -x + rd. \qquad (3)$$

Recall that if an inequality is multiplied by -1, the order of the inequality is reversed; for example, $2 < 3$, but $-3 < -2$. Multiplying the inequality in (3) by -1, we obtain the following inequality, which is equivalent to (2):

$$x - rd \leq \mu \leq x + rd. \qquad (4)$$

Thus we obtain

$$\text{pr}[x - rd \leq \mu \leq x + rd] \approx N(r) \qquad (5)$$

for sufficiently large values of n.

The statistician frequently phrases condition (5) in terms of his *confidence* that the actual mean μ falls within the interval $[x - rd, x + rd]$ of his empirical mean x. For example, if $N(r) = 0.5$, then he is 50% *confident* that μ falls in $[x - rd, x + rd]$, and, in general, the interval $[x - rd, x + rd]$ is his 100$N(r)$% *confidence interval for the actual mean* μ. The endpoints $x - rd$ and $x + rd$ of this interval are the 100$N(r)$% *confidence limits for the mean* μ.

   Suppose that in an attempt to find the mean number μ of heads for 100 flips of a certain possibly biased coin, a hundred flips are actually made and an empirical mean x is determined. The $100N(r)\%$ confidence interval $[x - rd, x + rd]$ for the actual mean cannot be determined, since the actual probability p of obtaining a head is not known, and of course, $d = \sqrt{npq}$ depends on p. However, we can estimate pq. If we let $t = \frac{1}{2} - p$, then $p = \frac{1}{2} - t$ and

$$pq = p(1 - p) = (\tfrac{1}{2} - t)[1 - (\tfrac{1}{2} - t)] = (\tfrac{1}{2} - t)(\tfrac{1}{2} + t)$$
$$= \tfrac{1}{4} - t^2.$$

Since $t^2 \geq 0$, we obtain

$$pq \leq \tfrac{1}{4}.$$

Therefore

$$d = \sqrt{npq} = \sqrt{n}\sqrt{pq} \leq \sqrt{n}\sqrt{\tfrac{1}{4}} = \frac{\sqrt{n}}{2}.$$

Of course, this estimate for d is most accurate if t is small, that is, if p is close to $\frac{1}{2}$. Since $d \leq \sqrt{n}/2$, the interval $[x - rd, x + rd]$ is a subset of

$$\left[x - r\frac{\sqrt{n}}{2}, x + r\frac{\sqrt{n}}{2} \right].$$

Thus this interval

$$\left[x - r\frac{\sqrt{n}}{2}, x + r\frac{\sqrt{n}}{2} \right]$$

is a safe $100N(r)\%$ confidence interval for μ. We summarize these results in a theorem.

   Theorem 9.1 (Confidence intervals for the mean μ). *For an independent-trials process in which n trials produce an empirical mean of x successes, a $100N(r)\%$ confidence interval for the actual mean μ is*

$$[x - rd, x + rd],$$

provided that n is sufficiently large. If d is not known, the interval

$$\left[x - r\frac{\sqrt{n}}{2}, x + r\frac{\sqrt{n}}{2} \right]$$

is also a $100N(r)\%$ confidence interval.

Example 9.2 Suppose that 100 flips of a possibly biased coin produce $x = 60$ heads. Let us find the 95% confidence interval for the actual mean number μ of heads for 100 flips of this coin.

   If $100N(r) = 95$, then $N(r) = 0.95$, so $PR[r] = 0.475$. The table in Fig. 8.3 shows that $r = 2$ will give a sufficiently large interval. Since we don't know the value of p, we take our crude (but safe) estimate

$$\left[x - r\frac{\sqrt{n}}{2}, x + r\frac{\sqrt{n}}{2} \right]$$

for this interval. Now

$$\frac{\sqrt{n}}{2} = \frac{\sqrt{100}}{2} = 5,$$

so we obtain a 95% confidence interval of $[60 - 2 \cdot 5, 60 + 2 \cdot 5] =$ [50, 70]. That is, we can have 95% confidence that $50 \leq \mu \leq 70$. ‖

9.2 Confidence Intervals for p

Since $p = \mu/n$, it is easy now for us to find confidence intervals for p. If

$$x - rd \leq \mu \leq x + rd$$

then

$$\frac{x}{n} - r\frac{d}{n} \leq \frac{\mu}{n} \leq \frac{x}{n} + r\frac{d}{n}.$$

Since

$$\frac{d}{n} = \frac{\sqrt{npq}}{n} = \frac{\sqrt{n}\sqrt{pq}}{(\sqrt{n})^2} = \frac{\sqrt{pq}}{\sqrt{n}},$$

our inequality becomes

$$p' - r\frac{\sqrt{pq}}{\sqrt{n}} \leq p \leq p' + r\frac{\sqrt{pq}}{\sqrt{n}}.$$

Thus a $100N(r)\%$ confidence interval for p is

$$\left[p' - r\frac{\sqrt{pq}}{\sqrt{n}}, p' + r\frac{\sqrt{pq}}{\sqrt{n}} \right].$$

Once again, we are often concerned with the case in which p' is determined experimentally and p is not known. The estimate $\sqrt{pq} \leq \frac{1}{2}$ is then appropriate. This leads to the $100N(r)\%$ confidence interval

$$\left[p' - \frac{r}{2\sqrt{n}}, p' + \frac{r}{2\sqrt{n}} \right].$$

We summarize these ideas in a theorem.

> **Theorem 9.2** (Confidence intervals for p). *For an independent-trials process in which n trials produce x successes, giving an empirical probability of $p' = x/n$, a $100N(r)\%$ confidence interval for the actual probability p is*
>
> $$\left[p' - r\frac{\sqrt{pq}}{\sqrt{n}}, p' + r\frac{\sqrt{pq}}{\sqrt{n}} \right]$$
>
> *for sufficiently large n. The estimate*
>
> $$\left[p' - \frac{r}{2\sqrt{n}}, p' + \frac{r}{2\sqrt{n}} \right]$$
>
> *is also a $100N(r)\%$ confidence interval for p.*

Example 9.3 Suppose 100 flips of a certain possibly biased coin produce 60 heads. We then obtain the empirical probability $p' = 60/100 = 0.6$ for a head on a single trial. As we saw in Example 9.2, a 95% confidence interval corresponds to $r = 2$. Thus we have 95% confidence that the actual value of p lies in the interval

$$\left[p' - \frac{r}{2\sqrt{n}}, p' + \frac{r}{2\sqrt{n}} \right] = \left[0.6 - \frac{2}{2\sqrt{100}}, 0.6 + \frac{2}{2\sqrt{100}} \right]$$
$$= [0.5, 0.7]. \parallel$$

Example 9.4 Suppose we wish to determine the fraction p of smokers in a certain large city. Let us determine how large a random sample needs to be chosen so that we can be 99% confident that the fraction p of smokers in the city differs from the fraction p' of smokers in the sample by at most 0.01.

We want to have 99% confidence that p lies in the interval $[p' - 0.01, p' + 0.01]$. For a 99% confidence interval, we have $N(r) = 0.99$, so $\mathrm{PR}[r] = 0.495$. We see from the table in Fig. 8.3 that $r = 2.6$ will suffice. By Theorem 9.2, it will suffice to have

$$\frac{r}{2\sqrt{n}} = \frac{2.6}{2\sqrt{n}} \le 0.01.$$

Therefore it suffices to have

$$\sqrt{n} \ge \frac{2.6 \cdot 100}{2} = 130 \quad \text{or} \quad n \ge 16,900.$$

Thus we can achieve our desired accuracy with 99% confidence by choosing a sample of 16,900 people. $\parallel$

It is interesting to note that for the situation described in Example 9.4, the confidence we can have in our empirical probability p' depends on only the size of the random sample selected for study, and is not affected by the size of the total population.

EXERCISES

9.1 From an experiment of 900 independent trials, an empirical probability of 0.61 is obtained for success on a single trial. Find a 90% confidence interval for the probability p of success on a single trial.

9.2 In an experiment of 10,000 independent trials, 7,000 successes are obtained. Find 80% confidence limits for the actual mean of the experiment.

9.3 The probability p of success for a certain independent-trials process is not known. However, it is known that p is fairly close to $\frac{1}{2}$. It is desired to establish by experimentation an empirical probability p' such that one can have 90% confidence that p' differs from p by at most 0.001. How many trials should be performed?

9.4 An election is to take place between candidates Smith and Jones. A poll of a random sample of 1600 voters shows that 52% of the voters sampled favor Smith. What percent confidence can the pollsters have that Smith will win the election, assuming that no voter changes his mind before the election?

9.5 Referring to Exercise 9.4, find 95% confidence limits for the fraction of the vote that Smith will receive, assuming that no voter changes his mind before the election.

9.6 A fair die is rolled 8000 times. Find 95% confidence limits for the number of 5's which appear.

9.7 A die is rolled 400 times, and on 140 rolls, either a 1 or a 2 is obtained. Can this be regarded as strong evidence that the die is not fair?

9.8 Answer Exercise 9.7 in the case that 160 rolls produce either a 1 or a 2.

9.9 A proprietor of a lunch counter knows from past experience that an average of two-fifths of his customers buy one hot dog each. (No one ever buys two hot dogs.) How many hot dogs should the proprietor have on hand if he wishes to be 99% confident of having enough hot dogs to serve his next 400 customers?

9.10 A drug company is testing the effectiveness of a vaccine for a certain disease. It is known from past tests that the mortality rate of mice infected with the disease is 60%. To test the effectiveness of the vaccine, the drug company vaccinates a random sample of 100 mice and then infects them with the disease. If 55 of these mice die from the disease, what percent confidence can the company have that their vaccine has some beneficial effect?

REFERENCES

1. C. B. ALLENDOERFER and C. O. OAKLEY, *Principles of Mathematics*, 2nd ed. New York: McGraw-Hill, 1963.

2. W. FELLER, *An Introduction to Probability Theory and Its Applications*, Vol. 1, 2nd ed. New York: Wiley, 1957.

3. J. G. KEMENY, J. L. SNELL, and G. L. THOMPSON, *Introduction to Finite Mathematics*, 2nd ed. Englewood Cliffs, N.J.: Prentice-Hall, 1966.

4. F. MOSTELLER, *Fifty Challenging Problems in Probability with Solutions.* Reading, Mass.: Addison-Wesley, 1965.

5. E. PARZEN, *Modern Probability Theory and Its Applications.* New York: Wiley, 1960.

6. F. J. SCHEID, *Elements of Finite Mathematics.* Reading, Mass.: Addison-Wesley, 1962.

3 | Real Analytic Geometry

The introduction of coordinates into geometry came in the seventeenth century with the work of René Descartes (1596–1650) and Pierre de Fermat (1601–1665). In Chapter 1 we saw how a line can be made into an endless ruler, using the real numbers, and we referred to such a ruler as a *number line*. We also saw in Chapter 1 that we could view $\mathbf{R}^2 = \mathbf{R} \times \mathbf{R}$ as the set of points in the Euclidean plane. In the present chapter, we shall consider these ideas and their natural extension in more detail.

Before we begin, let us take a moment to ask what the Euclidean line *is*. There is no physical object that you can point to and say, "*That* is the mathematicians' Euclidean line." The picture of a line which you draw with a pencil on a piece of paper is nothing more than a physical representation for you of something that exists only in your imagination. Of course, similar remarks can be made concerning the Euclidean plane.

There are two basic ways in which the mathematican attempts to put Euclidean geometry on a firm foundation.

APPROACH 1. In the *synthetic approach*, one assumes that there exists a set of undefined (or primitive) objects called *points* and another set of undefined objects called *lines*. One also has a rule, the *incidence relation*, which determines for each point and each line whether or not the point is incident with the line, i.e., lies on the line. In addition, one has *axioms* which are assumed to hold; these axioms concern points and lines and the incidence relation. For example, one axiom of Euclidean geometry, of any dimension, is the following:

Any two distinct points are incident with precisely one line.

One then deduces theorems from these axioms. (It is actually possible to introduce coordinates *by means of axioms*, and the study of the geometry can then proceed analytically, as we describe in Approach 2.)

APPROACH 2. In the *analytic approach* to Euclidean geometry, one *defines* a point on the Euclidean line to *be* a real number and a point of the Euclidean plane to *be* an element of $\mathbf{R} \times \mathbf{R}$. Thus in this approach, the Euclidean line *is* $\mathbf{R}$, and the Euclidean plane *is* $\mathbf{R}^2$. One then defines a line in $\mathbf{R}^2$ to be a certain type of subset of $\mathbf{R}^2$; for example, for a, b, $c \in \mathbf{R}$, where not both a and b are zero, the set

$$\{(x, y) \in \mathbf{R}^2 \mid ax + by + c = 0\}$$

is a line. This approach has the obvious advantage that the concepts of a point and a line are no longer undefined. However, one then immediately asks: "What is a real number? In particular, what is the real number 1?" Again a little thought will convince you that there is no physical object which *is* the real number 1, and indeed the black mark "1" is merely a convenient representation of something that exists only in your imagination. So we are in a sense no better off than with the synthetic approach. Eventually, the concept of a real number is reduced to the primitive, undefined concept of a set.

We shall be dealing chiefly with the analytic approach, and we shall depend on your geometric intuition to motivate our definitions. You developed this intuition in your high school, synthetic approach to Euclidean geometry.

1. DIMENSIONS ONE, TWO, AND THREE

1.1 Euclidean 1-, 2-, and 3-space

As we explained in the introduction, we define the **set of points of the Euclidean line (*1-space*)** to be the set **R**, and the **set of points of the Euclidean plane (*2-space*)** to be the set $\mathbf{R} \times \mathbf{R} = \mathbf{R}^2$. We give the helpful pictures for these two cases in Figs. 1.1 and 1.2, as we did in Chapter 1. The solid perpendicular lines of Fig. 1.2 are the **coordinate axes**. It is conventional to call the horizontal axis the "*x*-axis" and the vertical axis the "*y*-axis." However, since we shall proceed to the consideration of Euclidean *n*-space in Section 2,

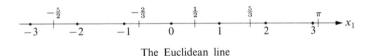

The Euclidean line

Fig. 1.1

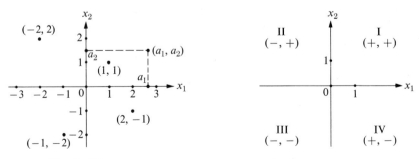

The Euclidean plane

Fig. 1.2

Fig. 1.3

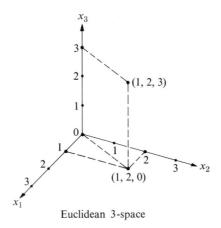

Euclidean 3-space

Fig. 1.4

we prefer to be more modern and call them the x_1-*axis* and the x_2-*axis*, as shown in Fig. 1.2. If we used different letters, we would run out of them after 26-space, but we will never run out of subscripts. The arrow on an axis indicates the positive direction. A point in the Euclidean plane is thus an ordered 2-tuple (a_1, a_2), and a_1 is the x_1-*coordinate* of the point, while a_2 is the x_2-*coordinate*. Figure 1.2 naturally divides $\mathbf{R}^2$ into four pieces or *quadrants*, according to the signs of the coordinates of the points. These quadrants are usually numbered as shown in Fig. 1.3. The point 0 in $\mathbf{R}$ and the point $(0, 0)$ in $\mathbf{R}^2$ are the *origins* of the line and the plane respectively.

We now define the *set of points of Euclidean space (3-space)* to be the set $\mathbf{R}^3$. To visualize this set "geometrically," we select as origin some point of our usual, synthetic 3-space picture, and imagine three coordinate axes, any two of which are perpendicular, through this point. In Fig. 1.4, we show just half of each of these x_1-, x_2-, and x_3-axes for clarity. It is not too easy to sketch a 3-space picture on a piece of paper.

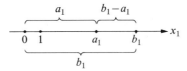

Fig. 1.5

1.2 Distance

Let us turn to the question of the *distance* between two points in a Euclidean geometry. First we consider the Euclidean line $\mathbf{R}$. For the points a_1 and b_1 shown in Fig. 1.5, we should surely define the distance between them to be $b_1 - a_1$. You can easily convince yourself that for two points a_1 and b_1 in $\mathbf{R}$, where $a_1 < b_1$, you may define the distance between them to be $b_1 - a_1$.

Example 1.1 The distance between -2 and 3 is $3 - (-2) = 5$, as indicated in Fig. 1.6. ‖

$$3-(-2)=5$$

$$\begin{array}{ccccccc} -3 & -2 & -1 & 0 & 1 & 2 & 3 \end{array} \quad x_1$$

Fig. 1.6

For *any* points a_1 and b_1 in **R**, we consider the distance between them to be either $a_1 - b_1$ or $b_1 - a_1$, whichever is nonnegative. This nonnegative value is often denoted by "$|a_1 - b_1|$", read "the **absolute value of** $a_1 - b_1$." Thus $|(-2) - 3| = |-5| = 5$. Another way of expressing this nonnegative difference between a_1 and b_1 is "$\sqrt{(a_1 - b_1)^2}$", where the square root symbol $\sqrt{\ }$ yields the *nonnegative* square root of a nonnegative number.

Example 1.2 For the points -2 and 3, we have

$$\sqrt{(3 - (-2))^2} = \sqrt{5^2} = \sqrt{25} = 5,$$

and also

$$\sqrt{((-2) - 3)^2} = \sqrt{(-5)^2} = \sqrt{25} = 5. \ \|$$

This square root expression will be more useful for us than the absolute value notation, for the square root formula for distance extends naturally to situations in 2-space and 3-space, as we shall soon see.

We want to make it plain that the preceding discussion of distance was just for *motivation* of the definition which now follows. In the analytic approach to the Euclidean line, one has a perfect right to simply define: "The **Euclidean line** is **R**, and for $a_1, b_1 \in$ **R**, the **distance between** a_1 **and** b_1 is $\sqrt{(a_1 - b_1)^2}$."

> **Definition.** For the Euclidean line **R**, the **distance between two points** a_1 **and** b_1 is $\sqrt{(a_1 - b_1)^2}$, where $\sqrt{\ }$ yields the nonnegative square root of a nonnegative number.

The motivation of the corresponding definition for the distance d between two points (a_1, a_2) and (b_1, b_2) in the Euclidean plane is provided by Fig. 1.7, together with the well-known Pythagorean theorem of synthetic plane geometry.

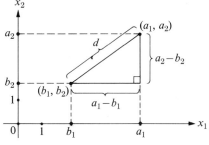

Fig. 1.7

Definition. For the Euclidean plane $\mathbf{R}^2$, the ***distance between two points*** (a_1, a_2) ***and*** (b_1, b_2) is

$$\sqrt{(a_1 - b_1)^2 + (a_2 - b_2)^2}.$$

Example 1.3 The distance between the points $(1, 3)$ and $(-2, 5)$ of $\mathbf{R}^2$ is

$$\sqrt{(1 - (-2))^2 + (3 - 5)^2} = \sqrt{3^2 + (-2)^2} = \sqrt{9 + 4} = \sqrt{13}. \;\|$$

Definition. The ***distance between two points*** (a_1, a_2, a_3) ***and*** (b_1, b_2, b_3) of Euclidean 3-space $\mathbf{R}^3$ is

$$\sqrt{(a_1 - b_1)^2 + (a_2 - b_2)^2 + (a_3 - b_3)^2}.$$

The motivation (not the proof!) of the preceding definition is given by Fig. 1.8. Here, in terms of high school geometry, the Pythagorean theorem gives us

$$d^2 = s^2 + (a_3 - b_3)^2$$

and

$$s^2 = (a_1 - b_1)^2 + (a_2 - b_2)^2.$$

Thus we must have

$$d^2 = (a_1 - b_1)^2 + (a_2 - b_2)^2 + (a_3 - b_3)^2.$$

Example 1.4 The distance between the points $(0, -3, 2)$ and $(4, 1, -1)$ of $\mathbf{R}^3$ is

$$\sqrt{(0 - 4)^2 + ((-3) - 1)^2 + (2 - (-1))^2} = \sqrt{16 + 16 + 9} = \sqrt{41}. \;\|$$

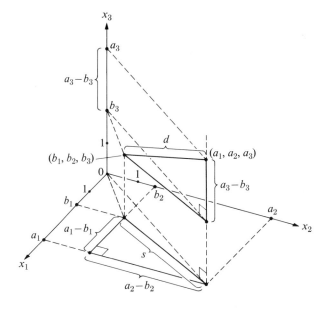

Fig. 1.8

1.3 Spheres and Balls

A *circle with center at* (a_1, a_2) *and a radius* $r > 0$ in the Euclidean plane is the set of all points in $\mathbf{R}^2$ of distance r from (a_1, a_2). From our distance formula, we see that a point (x_1, x_2) of $\mathbf{R}^2$ is on this circle if and only if

$$\sqrt{(x_1 - a_1)^2 + (x_2 - a_2)^2} = r,$$

or, squaring both sides of this equation, if and only if

$$(x_1 - a_1)^2 + (x_2 - a_2)^2 = r^2$$

(see Fig. 1.9).

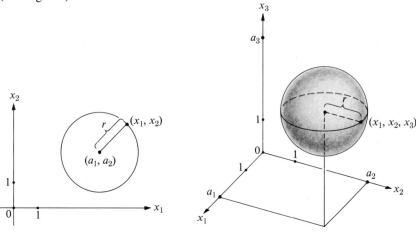

Fig. 1.9   Fig. 1.10

Example 1.5 In set notation, the circle in $\mathbf{R}^2$ with center at $(-1, 2)$ and radius of 3 is

$$\{(x_1, x_2) \in \mathbf{R}^2 \mid (x_1 + 1)^2 + (x_2 - 2)^2 = 9\}. \;\|$$

The *sphere in* $\mathbf{R}^3$ *with center* (a_1, a_2, a_3) *and radius* $r > 0$ consists of all points in $\mathbf{R}^3$ of distance r from (a_1, a_2, a_3). We see at once that a point (x_1, x_2, x_3) is on the sphere if and only if

$$(x_1 - a_1)^2 + (x_2 - a_2)^2 + (x_3 - a_3)^2 = r^2$$

(see Fig. 1.10).

Any mathematician would be led immediately by the preceding to consider, for the Euclidean line $\mathbf{R}$, all points $x_1 \in \mathbf{R}$ such that

$$(x_1 - a_1)^2 = r^2.$$

The set of such points contains just two points, $a_1 - r$ and $a_1 + r$, as shown in Fig. 1.11. What shall we call this subset of $\mathbf{R}$? Just as we dislike using different letters x, y, z for coordinate axes, so do we dislike in general inventing different names for similar things in different dimensions. We shall

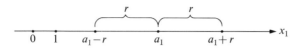

Fig. 1.11

be modern and call this set of two points in **R**, the circle in $\mathbf{R}^2$, and the sphere in $\mathbf{R}^3$ all "*spheres.*" Since the usual "sphere" in $\mathbf{R}^3$ is a two-dimensional object, we call it a "2-sphere," and we similarly call a circle a "1-sphere," for a curve is a one-dimensional object. A point is a zero-dimensional object.

Definition. The *0-sphere in* **R** *with center* a_1 *and radius* $r > 0$ is

$$\{x_1 \in \mathbf{R} \mid (x_1 - a_1)^2 = r^2\}.$$

The *1-sphere in* $\mathbf{R}^2$ *with center* (a_1, a_2) *and radius* $r > 0$ is

$$\{(x_1, x_2) \in \mathbf{R}^2 \mid (x_1 - a_1)^2 + (x_2 - a_2)^2 = r^2\}.$$

The *2-sphere in* $\mathbf{R}^3$ *with center* (a_1, a_2, a_3) *and radius* $r > 0$ is

$$\{(x_1, x_2, x_3) \in \mathbf{R}^3 \mid (x_1 - a_1)^2 + (x_2 - a_2)^2 + (x_3 - a_3)^2 = r^2\}.$$

For each of these sets, the defining equation is an ***equation of the sphere***.

Example 1.6 By the familiar process of *completing the square*, we can see that

$$S = \{(x_1, x_2) \in \mathbf{R}^2 \mid x_1^2 + x_2^2 - 2x_1 + 6x_2 - 6 = 0\}$$

is actually a 1-sphere, i.e., a circle. We have

$$
\begin{aligned}
x_1^2 + x_2^2 - 2x_1 + 6x_2 - 6 &= [x_1^2 - 2x_1] + [x_2^2 + 6x_2] - 6 \\
&= [(x_1 - 1)^2 - 1] + [(x_2 + 3)^2 - 9] - 6 \\
&= (x_1 - 1)^2 + (x_2 + 3)^2 - 16.
\end{aligned}
$$

Thus $x_1^2 + x_2^2 - 2x_1 + 6x_2 - 6 = 0$ is equivalent to

$$(x_1 - 1)^2 + (x_2 + 3)^2 = 16.$$

We see that the set S is a 1-sphere in $\mathbf{R}^2$ with center at $(1, -3)$ and radius 4. ‖

A 2-sphere in $\mathbf{R}^3$ bounds a *solid ball*, a 1-sphere in $\mathbf{R}^2$ encloses a *circular disk*, and a 0-sphere in **R** bounds a *line segment*. Again in order to avoid inventing different names for analogous configurations in different dimensions, the mathematician considers the solid ball to be a *3-ball*, the circular disk to be a *2-ball*, and the line segment to be a *1-ball*. An *n-ball of radius* $r > 0$ in a Euclidean space $\mathbf{R}^n$ consists of all points of distance $\leq r$ from a point which is the *center of the ball*.

Example 1.7 The 2-ball in $\mathbf{R}^2$ with radius 5 and center $(1, 2)$ is the set

$$\{(x_1, x_2) \in \mathbf{R}^2 \mid (x_1 - 1)^2 + (x_2 - 2)^2 \leq 25\}. \; ‖$$

EXERCISES

1.1 Give definitions for the following concepts, analogous to the definitions in this section. (We shall discuss these concepts in detail in the next section.)

a) Euclidean 4-space
b) The distance between two points in Euclidean 4-space
c) A 3-sphere in Euclidean 4-space
d) A 4-ball in Euclidean 4-space

1.2 The Euclidean plane $\mathbf{R}^2$ is divided into quadrants. Into how many pieces is the Euclidean 3-space $\mathbf{R}^3$ divided according to the possible signs of the coordinates?

1.3 In each case, find the distance between the given points of $\mathbf{R}$.

a) 2 and 5   b) -1 and 4   c) $\sqrt{2}$ and $-\pi$

1.4 In each case, find the distance between the given points of $\mathbf{R}^2$.

a) $(3, -1)$ and $(0, 3)$ b) $(0, 0)$ and $(3, \pi)$
c) $(\sqrt{2}, 3)$ and $(0, \frac{1}{2})$

1.5 In each case, find the distance between the given points of $\mathbf{R}^3$.

a) $(2, 1, -3)$ and $(1, 1, 1)$ b) $(0, 2, \frac{3}{2})$ and $(-3, 1, -\frac{1}{2})$

1.6 Sketch each of the following subsets of $\mathbf{R}^2$ in a suitable figure.

a) $\{(x_1, x_2) \in \mathbf{R}^2 \mid x_1 = 2\}$ b) $\{(x_1, x_2) \in \mathbf{R}^2 \mid x_2 = -2\}$
c) $\{(x_1, x_2) \in \mathbf{R}^2 \mid x_1 = x_2\}$ d) $\{(x_1, x_2) \in \mathbf{R}^2 \mid x_1^2 + x_2^2 = 1\}$

1.7 Sketch each of the following subsets of $\mathbf{R}^3$ in a suitable figure.

a) $\{(x_1, x_2, x_3) \in \mathbf{R}^3 \mid x_1 = 2\}$ b) $\{(x_1, x_2, x_3) \in \mathbf{R}^3 \mid x_3 = 3\}$
c) $\{(x_1, x_2, x_3) \in \mathbf{R}^3 \mid x_1 = x_2\}$

1.8 Although we have not defined a *line* in $\mathbf{R}^2$, use your geometric intuition to find the following points. Draw figures.

a) The point of $\mathbf{R}^2$ such that the line segment joining it to $(2, -1)$ has the x_1-axis as perpendicular bisector
b) The point in $\mathbf{R}^2$ such that the line segment joining it to $(-3, 2)$ has the x_2-axis as perpendicular bisector
c) The point in $\mathbf{R}^2$ such that the line segment joining it to $(-1, 3)$ has the origin as midpoint
d) The point in $\mathbf{R}^2$ such that the line segment joining it to $(2, -4)$ has $(2, 1)$ as midpoint

1.9 Although we have defined neither a *line* nor a *plane* in $\mathbf{R}^3$, use your geometric intuition to find the following points.

a) The point of $\mathbf{R}^3$ such that the line segment joining it to $(-2, 1, -4)$ is bisected by and is perpendicular to the plane $\{(0, x_2, x_3) \mid x_2, x_3 \in \mathbf{R}\}$
b) The point in $\mathbf{R}^3$ such that the line segment joining it to $(-1, \pi, \sqrt{2})$ has the origin as midpoint
c) The point in $\mathbf{R}^3$ such that the line segment joining it to $(-1, 4, -3)$ has as midpoint $(-1, 2, -3)$
d) The point in the plane $\{(x_1, 2, x_3) \mid x_1, x_3 \in \mathbf{R}\}$ which is closest to the point $(-1, -5, 2)$

1.10 Find an equation for each of the following spheres.

a) The 1-sphere with center $(0, 1)$ and radius 5
b) The 2-sphere with center $(3, -1, -2)$ and radius 4
c) The 0-sphere with center 7 and radius 4
d) The 0-sphere $\{-2, 8\}$

1.11 Sketch each of the following subsets of $\mathbf{R}^3$ in a suitable figure.

a) $\{(x_1, x_2, x_3) \in \mathbf{R}^3 \mid x_2^2 = 1\}$
b) $\{(x_1, x_2, x_3) \in \mathbf{R}^3 \mid x_1^2 + x_2^2 = 1\}$
c) $\{(x_1, x_2, x_3) \in \mathbf{R}^3 \mid x_1^2 + x_2^2 + x_3^3 = 1\}$
d) $\{(x_1, x_2, x_3) \in \mathbf{R}^3 \mid x_1 = x_2 = x_3\}$

1.12 Express the 2-ball in $\mathbf{R}^2$ with center $(-1, 4)$ and radius 3 in set notation.

1.13 Express the 3-ball in $\mathbf{R}^3$ with center $(1, 0, -3)$ and radius 2 in set notation.

1.14 By completing the square, show that the set

$$\{(x_1, x_2) \in \mathbf{R}^2 \mid x_1^2 + x_2^2 - 4x_1 + 8x_2 - 5 = 0\}$$

is a 1-sphere. Find its center and radius.

1.15 By completing the square, show that the equation

$$x_1^2 + x_2^2 + x_3^2 + 2x_2 - 6x_3 - 6 = 0$$

is the equation of a 2-sphere. Find its center and radius.

2. HIGHER-DIMENSIONAL SPACES

2.1 Euclidean n-space, n-spheres, and n-balls

Often when the instructor starts to talk about Euclidean 4-space, some student will say, "But there *is* no such thing." Remember that in the physical sense, there *is* no Euclidean 1-, 2-, or 3-space either. They all exist only in our imagination. Now surely it is no more of a feat for us to imagine an ordered 4-tuple (a_1, a_2, a_3, a_4) of real numbers than to imagine the ordered 3-tuple (a_1, a_2, a_3). That is, the existence of $\mathbf{R}^4$ is just as real as the existence of $\mathbf{R}$, $\mathbf{R}^2$, and $\mathbf{R}^3$. You should be able to give our next definitions by yourself with no trouble.

> **Definition.** For $n \in \mathbf{Z}^+$, the **set of points in Euclidean n-space** is the set $\mathbf{R}^n$. The **distance between two points** $(a_1, \ldots, a_n)$ **and** $(b_1, \ldots, b_n)$ of $\mathbf{R}^n$ is
> $$\sqrt{(a_1 - b_1)^2 + \cdots + (a_n - b_n)^2}.$$

> **Definition.** The **n-sphere in $\mathbf{R}^{n+1}$ with center** $(a_1, \ldots, a_n)$ **and radius** $r > 0$ is
> $$\{(x_1, \ldots, x_{n+1}) \in \mathbf{R}^{n+1} \mid (x_1 - a_1)^2 + \cdots + (x_{n+1} - a_{n+1})^2 = r^2\}.$$

> **Definition.** The **n-ball in $\mathbf{R}^n$ with center** $(a_1, \ldots, a_n)$ **and radius** $r > 0$ is
> $$\{(x_1, \ldots, x_n) \in \mathbf{R}^n \mid (x_1 - a_1)^2 + \cdots + (x_n - a_n)^2 \leq r^2\}.$$

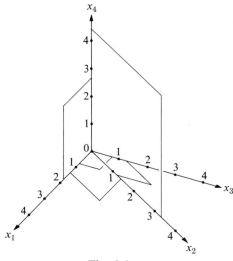

Fig. 2.1

It is easy to draw helpful pictures for **R** and **R**² on a piece of paper. It is harder to draw a picture for **R**³, and harder still to draw a helpful picture for **R**⁴, on paper, but let us attempt it. Through one point, the origin, we need four coordinate axes, any two of which are perpendicular. In Fig. 2.1, we have just drawn half of each axis for clarity. We hope that if you cover up any axis with your pencil, you can at least visualize how any two of the three remaining axes are perpendicular. Thus any two of these axes are perpendicular.

*2.2 The Amount of Room in **R**ⁿ

For the rest of this section we shall attempt to strengthen your geometric intuition regarding Euclidean 4-space **R**⁴. One way to gain insight into the geometric properties of **R**⁴ is to imagine that you are an inhabitant of one of the Euclidean spaces **R** or **R**², and see what things you *could not do* which an inhabitant of a higher-dimensional space *could do*.

Let us start with **R**. Suppose that you are a one-dimensional bug living in a one-dimensional space, the line **R**. Let us suppose that you occupy the line **R** from −1 to 1 (see Fig. 2.2). Now as a one-dimensional bug, with only

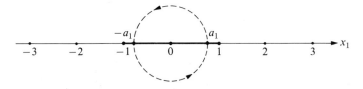

Fig. 2.2

length and no width or thickness, *you simply can't imagine that your line world is part of a plane.* Thus, assuming that you are facing the positive direction on the x_1-axis, you simply can't imagine what it would mean to "turn around" and face the other direction. We express this mathematically by saying that you cannot imagine how to change your *orientation.* Now if only you could see that your line was part of a plane, you would see that, *with your body kept rigid,* you could turn around in this plane until you were facing the other way in your line world. The point 0 of your body could remain fixed, and the point a_1 of your body could travel the top half of the dashed circle in Fig. 2.2, while the point $-a_1$ could travel the bottom half of this circle. It wouldn't hurt you at all.

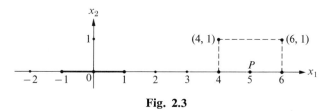

Fig. 2.3

For a further illustration using a one-dimensional bug, suppose that, as a one-dimensional bug in **R**, you are crawling happily along in the positive direction when someone puts a point of dirt P in front of you, say at the point 5 (see Fig. 2.3). Now you can see this piece of dirt; it is in your world; and you realize that there is more of your world beyond it, but you can't get to the rest of your world until it is removed. If only you could see that your world was part of a plane, you could go up close to P, and *go around it,* following the dashed path shown in Fig. 2.3. Then you could continue merrily on your way. A path constructed of a series of straight line segments is a ***polygonal path.*** It is true that, in getting from 1 to 6, your head would travel a total distance of seven units for the polygonal path in Fig. 2.3, rather than the five units for the straight line path. Clearly, a shorter polygonal path could be used, say from

$$(1, 0) \text{ to } (4, 0) \text{ to } (4, \tfrac{1}{2}) \text{ to } (6, \tfrac{1}{2}) \text{ to } (6, 0).$$

The length of this path is only six units. Of course, you could never quite achieve the straight line distance of five units, for that path is blocked by your piece of dirt.

Let us now proceed to analogous considerations for a two-dimensional bug. Suppose that you are a two-dimensional bug living in **R²** (see Fig. 2.4). As a two-dimensional bug, you can't imagine that your plane world is a part of **R³**. You have an idea of front versus back, i.e., x_1-direction, and of left versus right, i.e., x_2-direction, but you have no conception of up and down, i.e., x_3-direction. You can't imagine ***reversing your orientation,*** which always means changing one coordinate direction while keeping the others fixed. For

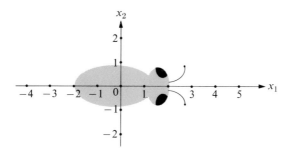

Fig. 2.4

example, you can't imagine reversing your left and right sides, while still facing the same way, that is, keeping the same directions for front and back. Now if only you could see that your plane was part of $\mathbf{R}^3$, you would see that, *with your body kept rigid*, you could "turn over," so that your left and right sides would be reversed. It wouldn't hurt you at all. The part of your body on the x_1-axis (see Fig. 2.4) could remain stationary, while a part of your body not on the x_1-axis could travel on a semicircle, always keeping a constant distance from the x_1-axis.

Suppose now that, as a two-dimensional bug, you are crawling happily along the x_1-axis in the positive direction when someone places a line L (a one-dimensional space) in your plane world, cutting the x_1-axis at 5 (see Fig. 2.5). That line cuts you off from the rest of your plane world. You can't go around it; it extends without end in both directions. You would have to cut a gate in it, or have it removed, to get to the other side. Now if only you could see that your plane world was part of $\mathbf{R}^3$, you could travel the polygonal path

$$(2, 0, 0) \text{ to } (4, 0, 0) \text{ to } (4, 0, 1) \text{ to } (6, 0, 1) \text{ to } (6, 0, 0)$$

(see Fig. 2.5), hopping over this line back into your plane world, and continue merrily on your way. This polygonal path from $(2, 0, 0)$ to $(6, 0, 0)$ of course has a length of six units, rather than the four-unit length of the straight line path on the x_1-axis.

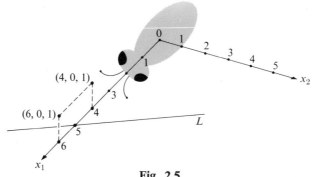

Fig. 2.5

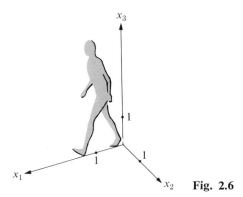

Fig. 2.6

Finally, after all this preparation, let us suppose that you are a three-dimensional bug living in $\mathbf{R}^3$ (see Fig. 2.6), and are happily walking along the x_1-axis in the positive direction. You can distinguish front and back, i.e., x_1-direction, left and right, i.e., x_2-direction, and up and down, i.e., x_3-direction. But you simply can't imagine the x_4-direction. You also can't imagine reversing your orientation, for example, interchanging your left and right sides (arms, legs, lungs, etc.) while still facing the positive direction on the x_1-axis and still pointing your head up the x_3-axis. (This amounts to reversing your x_2-direction, keeping the x_1- and x_3-directions fixed.) Now if only you could see that your $\mathbf{R}^3$ is part of $\mathbf{R}^4$, you would realize that, *with your body kept rigid*, you could "reverse" yourself, so that your left and right sides would be interchanged. It wouldn't hurt you at all. Again, each point of your body would simply travel a semicircular path.

Suppose that while you are walking happily along the x_1-axis, someone places a plane (a 2-space) in your 3-space, perhaps meeting the x_1-axis at 5 (see Fig. 2.7); that plane blocks your path. You can see it; you know that there is more of your three-dimensional world beyond it, but you can't get to the other side of it without chopping a hole through it or having it removed. (It extends infinitely, so you can't go around it.) Now if only you could see that your 3-space $\mathbf{R}^3$ was part of $\mathbf{R}^4$, say the part

$$\{(x_1, x_2, x_3, x_4) \in \mathbf{R}^4 \mid x_4 = 0\},$$

you would see that you could travel the polygonal path

$(1, 0, 0, 0)$ to $(4, 0, 0, 0)$ to $(4, 0, 0, 1)$ to $(6, 0, 0, 1)$ to $(6, 0, 0, 0)$,

come back into your 3-space on the other side of the plane, and continue merrily on your way.

Thus in 4-space $\mathbf{R}^4$, it is possible to get from one side of the surface to the other side without going through the surface.

We hope that your geometric intuition for higher-dimensional spaces has been strengthened. Some further help is provided in the exercises. For

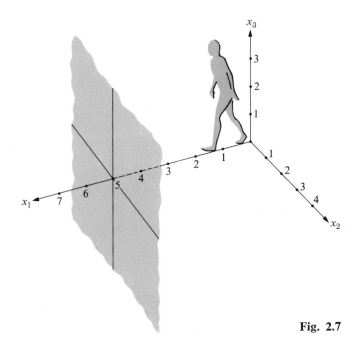

Fig. 2.7

those intrigued by the preceding discussion, we strongly recommend the excerpt from E. A. Abbott's *Flatland* (New York: Barnes and Noble, 1963) with the accompanying commentary in Newman [2, Vol. 4, page 2383].

Thus R^2 has more room in which to maneuver than R, R^3 more room than R^2, R^4 more room than R^3, etc. It is even possible to define a space of *ordered infinite-tuples*; this space, **Hilbert space**, contains the spaces R^n for all positive integers n. Hilbert space is very roomy indeed. In fact, the jingle

> Hilbert space! Hilbert space!
> Lots more room in Hilbert space!,

set to the tune of a well-known radio soap commercial of the 1940's, has enjoyed some popularity.

*2.3 An Application

In closing, we describe an application of R^4 to topology. Topology is one of the most active branches of geometry at the present time.

Note that a 2-sphere has two sides, an "inside" and an "outside." A water glass also has an inside and an outside; a fly on a water glass can't crawl from one side to the other without crossing the rim or *edge* of the glass. A 2-sphere has no edges, and the fly can't crawl from the outside to the inside in any fashion. Does there exist a surface on which a fly can crawl "from one side to the other" *without crossing an edge?* The *Möbius strip*, formed by joining the ends of a strip of paper with a half twist, as shown in

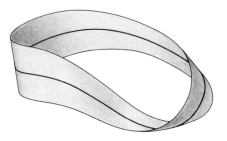

Fig. 2.8

Fig. 2.8, is such a surface. If a fly crawls "around it," he winds up on the "other side." In this sense, a Möbius strip is really a one-sided surface. However, a Möbius strip has an edge. Does there exist such a one-sided surface *without edges?* The answer is *yes*, but such a surface does not exist in the world $\mathbf{R}^3$. One must go to $\mathbf{R}^4$. We give the most famous example of such a surface, the **Klein bottle**. The formation of a Klein bottle is shown in Fig. 2.9. Imagine that you start with a cylinder made of soft, stretchable

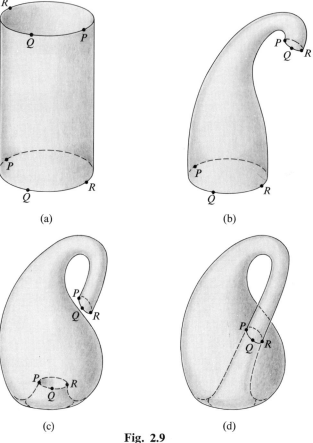

(a)   (b)

(c)   (d)

Fig. 2.9

rubber, as shown in Fig. 2.9(a). The job is to bring the circular ends together so that the two P's, the two Q's, and the two R's come together. This can't be done in $\mathbf{R}^3$. Parts (b), (c), and (d) of Fig. 2.9 show how this is accomplished by stretching and deforming the cylinder. To form Fig. 2.9(d) from (c), one must bring the elongated "neck" shown in (b) around and "through" itself without allowing it to intersect itself. This requires going from one side of a surface to the other without going through it, and we have said that this can be done in $\mathbf{R}^4$. The completed Klein bottle is shown in the Fig. 2.9(d). If you imagine that you are a fly crawling along the "neck," you can get to the "other side." Thus a Klein bottle really has only one side. If you try to paint "one side," you will wind up painting the whole thing. A Klein bottle has no edges; that is, it is a *closed surface*.

EXERCISES

*2.1 A one-dimensional bug living in $\mathbf{R}$ wants to crawl from -3 to 3, but a piece of dirt P at the origin blocks his path.

a) Describe a polygonal path, as we did in the text, which the bug can travel in $\mathbf{R}^2$ to avoid the piece of dirt P. How long is your path?

b) Describe a polygonal path as in part (a), but make the total length of the path exactly 6.1 units.

*2.2 A two-dimensional bug living in $\mathbf{R}^2$ at the origin is suddenly enclosed by the 1-sphere with equation $x_1^2 + x_2^2 = 1$.

a) Describe a polygonal path in $\mathbf{R}^3$ by which the bug can escape to the point $(1, 1)$ outside the 1-sphere. What is the length of your path?

b) Answer part (a), but make the total length of the path exactly two units.

*2.3 A three-dimensional bug living in $\mathbf{R}^3$ at the origin is suddenly trapped by the 2-sphere with equation $x_1^2 + x_2^2 + x_3^3 = 1$.

a) Describe a polygonal path in $\mathbf{R}^4$ by which he could escape to the point $(1, -1, 1)$, outside the 2-sphere. What is the length of your path?

b) Answer part (a), but make the total length of the path exactly two units.

*2.4 Argue that it should be possible for a three-dimensional man living in $\mathbf{R}^3$ to face "north" with his right side toward the "east" while standing on his head, if he can utilize $\mathbf{R}^4$.

*2.5 Consider a man's two-dimensional silhouette living in $\mathbf{R}^2$. Show that, when regarded by a superior dweller in $\mathbf{R}^3$, the silhouette's left and right gloves are indistinguishable (see Fig. 2.10).

Left glove   Right glove   **Fig. 2.10**

*2.6 Following the idea of Exercise 2.5, argue that if we hire a four-dimensional manufacturer in $\mathbf{R}^4$ to make our gloves, he would only need to make one shape glove. (Do *not* confuse this with turning a glove inside out. The gloves could be made of iron.)

*2.7 Describe some tricks that a two-dimensional magician living in $\mathbf{R}^2$ could do to amaze his two-dimensional fellow creatures, assuming that the magician could utilize $\mathbf{R}^3$.

*2.8 Repeat Exercise 2.7 for a three-dimensional magician living in $\mathbf{R}^3$ who can utilize $\mathbf{R}^4$.

*2.9 Devise a polygonal path in $\mathbf{R}^{11}$ by means of which a ten-dimensional bug living at the origin in $\mathbf{R}^{10}$ could escape from the surrounding 9-sphere with equation $x_1^2 + \cdots + x_{10}^2 = 1$ to the point $(1, -1, 1, -1, 1, -1, 1, -1, 1, -1)$. What is the length of your path?

*2.10 Consider a two-dimensional bug having as his world a Klein bottle surface. Argue that the bug can reverse his orientation, for example, change his left and right sides, by crawling along a suitable path *in his world*.

*2.11 If a 1-sphere in $\mathbf{R}^2$ is intersected with a line (an $\mathbf{R}^1$) lying in $\mathbf{R}^2$ and not tangent to the 1-sphere, then a set of two points, that is, a 0-sphere, results. If a 2-sphere in $\mathbf{R}^3$ is intersected with a plane (an $\mathbf{R}^2$) lying in $\mathbf{R}^3$ and not tangent to the 2-sphere, a circle, that is, a 1-sphere, results. If a 2-sphere is intersected with a nontangent line (an $\mathbf{R}^1$) lying in $\mathbf{R}^3$, then a set of two points, a 0-sphere, results.

a) Extend these ideas one step further to a 3-sphere in $\mathbf{R}^4$.
b) If a 4-sphere in $\mathbf{R}^5$ is intersected with a "flat three-dimensional piece" (an $\mathbf{R}^3$) lying in $\mathbf{R}^5$ which is not tangent to the 4-sphere, what results?
c) Generalize (b) to an n-sphere in $\mathbf{R}^{n+1}$ intersected with a "flat s-dimensional piece" (an $\mathbf{R}^s$) of $\mathbf{R}^{n+1}$ which is not tangent to the n-sphere.

*2.12 The Euclidean line world of a one-dimensional bug is part of the Euclidean plane world of a two-dimensional bug. The two-dimensional bug crawls across the line world of the one-dimensional bug. What does the one-dimensional bug see?

*2.13 The Euclidean plane world of a two-dimensional bug is part of the Euclidean 3-space world of a three-dimensional bug. The three-dimensional bug crawls through the plane world of the two-dimensional bug. What does the two-dimensional bug see?

*2.14 The Euclidean 3-space world of a three-dimensional bug is part of the Euclidean 4-space world of a four-dimensional bug. The four-dimensional bug crawls "through" the 3-space world of the three-dimensional bug. What does the three-dimensional bug see?

3. VECTORS

3.1 The Notion of a Vector

Real analytic geometry has many applications to the solution of physical problems. For example, suppose you are standing on a plane surface and observing another object on the surface. It is natural for you to ask, "How far away from me is that object, and in what direction?" The question of the

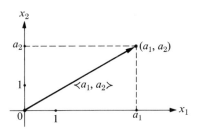

Fig. 3.1

distance from you to the object is most easily described in terms of some scale measuring distance, and direction is often described in terms of some known direction such as "north." If you imagine that the plane surface on which you are standing is a model of $\mathbf{R}^2$, with the origin located at your position and "north" in the direction of the positive x_2-axis, then both the distance and the direction from you to the object which you are observing are determined by the coordinates of the point where it is located. Let us imagine an arrow drawn from you to the object (see Fig. 3.1). This arrow represents both the direction (the arrowhead) and the distance (the length of the arrow) in which you are interested. Physicists call a quantity representing direction and magnitude a "*vector*". While the point (a_1, a_2) itself does determine both the direction and the distance, we shall introduce the notation "$\prec a_1, a_2 \succ$" for the vector (see Fig. 3.1). Many texts use boldface type for vectors, but we feel that this notation is better, since you can write it easily in your work. The *length* of the vector $\prec a_1, a_2 \succ$ is then $\sqrt{a_1^2 + a_2^2}$. We give the obvious generalization of these ideas to $\mathbf{R}^n$ in a definition.

> **Definition.** A ***vector in*** $\mathbf{R}^n$ is an ordered n-tuple $\prec a_1, a_2, \ldots, a_n \succ$ of real numbers. The ***length of the vector***, denoted by "$\|\prec a_1, a_2, \ldots, a_n \succ\|$", is $\sqrt{a_1^2 + a_2^2 + \cdots + a_n^2}$. The number a_i is the ***ith component of the vector***. Any vector of length 1 is a ***unit vector***.

Example 3.1 The vectors

$$\prec 0, 1 \succ \quad \text{and} \quad \left\langle \frac{1}{2}, \frac{\sqrt{3}}{2} \right\rangle$$

are both unit vectors in $\mathbf{R}^2$. ‖

3.2 Scalar Multiplication

When studying motion, a physicist may use a vector to represent a *force*. Suppose, for example, that you are moving some object by pushing it. The motion of the object is influenced by the *direction* in which you are pushing and how *hard* you are pushing. Thus the force with which you are pushing can be conveniently represented by a vector, which we think of as having the *direction* in which you are pushing and a *length* representing how hard you are pushing. If you double the force with which you push, the force vector doubles in length.

There are two basic operations concerning vectors which we wish to introduce, and both operations are nicely motivated by considering the physicist's interpretation of a vector as a force. We shall use Greek letters α, β, γ, δ for vectors. When dealing with vectors, one often refers to a number $r \in \mathbf{R}$ as a *scalar*, to distinguish it from a vector.

First, suppose you double the force with which you are pushing an object. If α is the original force vector, it is natural to call the new force vector 2α. For a vector α in $\mathbf{R}^n$ and a scalar $r \in \mathbf{R}$, we shall define in general a *scalar multiplication of α by r*. We shall then show that the resulting vector $r\alpha$ has the direction of α for $r > 0$, and is of length $|r| \cdot \|\alpha\|$, where $|r|$ is either r or $-r$, whichever is nonnegative.

Definition. Let $\alpha = \langle a_1, a_2, \ldots, a_n \rangle$ and let $r \in \mathbf{R}$. The ***product*** $r\alpha$ ***of the scalar*** r ***and the vector*** α is the vector $\langle ra_1, ra_2, \ldots, ra_n \rangle$.

Theorem 3.1 *If* $\alpha = \langle a_1, a_2, \ldots, a_n \rangle$ *and* $r \in \mathbf{R}$, *then* $\|r\alpha\| = |r| \cdot \|\alpha\|$, *where* $|r|$ *is either* r *or* $-r$, *whichever is nonnegative.*

Proof. Let $r \in \mathbf{R}$ and $\alpha = \langle a_1, a_2, \ldots, a_n \rangle$. Then we have

$$\|\langle ra_1, ra_2, \ldots, ra_n \rangle\| = \sqrt{(ra_1)^2 + (ra_2)^2 + \cdots + (ra_n)^2}$$
$$= \sqrt{r^2(a_1^2 + a_2^2 + \cdots + a_n^2)}$$
$$= \sqrt{r^2} \cdot \sqrt{a_1^2 + a_2^2 + \cdots + a_n^2}$$
$$= \sqrt{r^2} \cdot \|\langle a_1, a_2, \ldots, a_n \rangle\|$$
$$= |r| \cdot \|\langle a_1, a_2, \ldots, a_n \rangle\|,$$

where $|r| = \sqrt{r^2}$, that is, $|r|$ is either r or $-r$, whichever is nonnegative. ∎

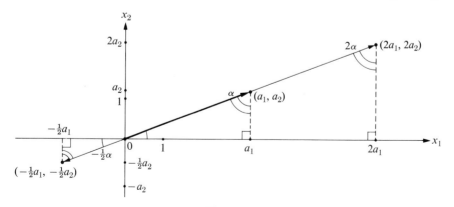

Fig. 3.2

The similar triangles shown in the $\mathbf{R}^2$-picture in Fig. 3.2 illustrate that we should consider $r\alpha$ to have the same direction as α if $r > 0$, and the opposite direction if $r < 0$.

Definition. Two vectors α and β in $\mathbf{R}^n$ are **parallel** if $\beta = r\alpha$ for some $r \in \mathbf{R}$.

Thus $r\alpha$ is a vector parallel to α of length $|r| \cdot \|\alpha\|$ and having the **same direction as** α if $r > 0$, and the **opposite direction** if $r < 0$.

Example 3.2 The vectors $\langle 1, -3 \rangle$ and $\langle 2, -6 \rangle$ in $\mathbf{R}^2$ are parallel and have the same direction, for $\langle 2, -6 \rangle = 2(\langle 1, -3 \rangle)$. However, $\langle 1, -3 \rangle$ and $\langle 2, -7 \rangle$ are *not* parallel. ‖

Example 3.3 Let $\|\alpha\| = 5$. Then

$$\|3\alpha\| = |3| \cdot \|\alpha\| = 3 \cdot 5 = 15,$$

and 3α has the same direction as α. However, -7α has the opposite direction to α, and

$$\|-7\alpha\| = |-7| \cdot \|\alpha\| = 7 \cdot 5 = 35. \; \|$$

3.3 Vector Addition

Suppose that two people are pushing an object with forces which correspond to force vectors α and β, as shown in Fig. 3.3. It can be shown that the motion of the object due to these combined forces is the same as the motion that would result if only one person were pushing with the force having as vector the diagonal of the parallelogram with arrow vectors α and β as adjacent sides (see Fig. 3.3). It is natural to call the vector on the diagonal the "*sum* of α and β." As indicated in Fig. 3.4, this vector $\alpha + \beta$ can be found by adding the corresponding components of the two vectors α and β. The difference of two vectors is defined similarly.

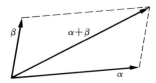

Fig. 3.3

Definition. Let $\alpha = \langle a_1, a_2, \ldots, a_n \rangle$ and $\beta = \langle b_1, b_2, \ldots, b_n \rangle$ both be vectors in $\mathbf{R}^n$ for the same value of n. The **sum** $\alpha + \beta$ **of** α **and** β is the vector in $\mathbf{R}^n$ given by

$$\alpha + \beta = \langle a_1 + b_1, a_2 + b_2, \ldots, a_n + b_n \rangle,$$

and the **difference** $\alpha - \beta$ **of** α **and** β is the vector given by

$$\alpha - \beta = \langle a_1 - b_1, a_2 - b_2, \ldots, a_n - b_n \rangle.$$

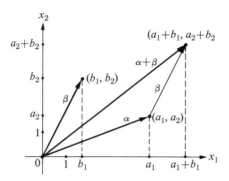

Fig. 3.4

Thus to compute sums or differences of vectors, we simply add or sub-tract corresponding components of the vectors. *Note that both vectors must be n-tuples for the same value of n.*

Example 3.4 We have

$$\langle -3, 2, 0 \rangle + \langle 5, -4, -1 \rangle = \langle 2, -2, -1 \rangle,$$

while

$$\langle -3, 2, 0 \rangle - \langle 5, -4, -1 \rangle = \langle -8, 6, 1 \rangle. \;\|$$

Note in Fig. 3.4 that we can arrive at the "tip of $\alpha + \beta$" by starting at the "tip of α" and going in the direction of β for the distance given by β. Since the arrow from the "tip of α" to the "tip of $\alpha + \beta$" has the same direction and length as the arrow from the origin to (b_1, b_2), we label this arrow β also in Fig. 3.4. Remember that we are intuitively thinking of a vector as a quantity with *magnitude* and *direction*, and both these arrows have the same *length* and the same *direction*. Of course, these pictures are just visual aids.

Clearly, by our definitions, $\alpha - \beta = \alpha + (-1)\beta$, as you would expect. Since $\alpha - \beta$ is the vector which when added to β gives α, that is,

$$(\alpha - \beta) + \beta = \alpha,$$

we can think of the length and direction of $\alpha - \beta$ as being given by an arrow going from the "tip of β" to the "tip of α" (see Fig. 3.5).

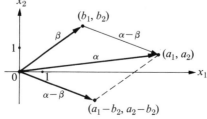

Fig. 3.5

You can easily check from our definitions that the following formal properties are true for any vectors α, β, γ in $\mathbf{R}^n$ and scalars r, $s \in \mathbf{R}$. The student who knows a little abstract algebra realizes that these properties allow us to use the usual techniques of algebra for computation with vectors.

1. $(\alpha + \beta) + \gamma = \alpha + (\beta + \gamma)$
2. $\alpha + \beta = \beta + \alpha$
3. $r(s\alpha) = (rs)\alpha$
4. $(r + s)\alpha = r\alpha + s\alpha$
5. $r(\alpha + \beta) = r\alpha + r\beta$

The *zero vector in* $\mathbf{R}^n$ is $\langle 0, 0, \ldots, 0 \rangle$. Of course, it has length 0, and we shall see in a moment that it is convenient to consider the zero vector to have all directions.

3.4 Orthogonal Vectors

It is very important for us to know when two directions are *perpendicular* or, to use more modern terminology, *orthogonal*. An elegant definition of this concept is easily given in terms of vectors, and this is one of the reasons we introduced vectors in this text. Let us motivate our definition with a picture and a synthetic argument.

As you know from high school geometry, in $\mathbf{R}^3$, and also in $\mathbf{R}^n$, any three points, not all on the same line, determine a plane. (Note that we have not yet *defined* a line or a plane. We shall do this in the following sections.) We imagine that Fig. 3.6 gives a picture of such a plane determined by three points, $(0, 0, \ldots, 0)$, $(a_1, a_2, \ldots, a_n)$, and $(b_1, b_2, \ldots, b_n)$ in $\mathbf{R}^n$. It would be reasonable to define the vectors α and β of Fig. 3.6 to be orthogonal (perpendicular) if and only if the Pythagorean relation holds, that is, if and only if

$$\|\alpha\|^2 + \|\beta\|^2 = d^2 = \|\alpha - \beta\|^2.$$

By definition of the length of a vector, we see that this is true when

$$(a_1^2 + a_2^2 + \cdots + a_n^2) + (b_1^2 + b_2^2 + \cdots + b_n^2)$$

$$= (a_1 - b_1)^2 + (a_2 - b_2)^2 + \cdots + (a_n - b_n)^2. \quad (1)$$

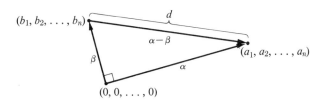

Fig. 3.6

Now it is easy to see that

$$(a_1 - b_1)^2 + (a_2 - b_2)^2 + \cdots + (a_n - b_n)^2$$
$$= (a_1{}^2 + a_2{}^2 + \cdots + a_n{}^2) + (b_1{}^2 + b_2{}^2 + \cdots + b_n{}^2)$$
$$- 2(a_1 b_1 + a_2 b_2 + \cdots + a_n b_n)$$

Substituting in (1) and canceling

$$a_1{}^2 + a_2{}^2 + \cdots + a_n{}^2 \quad \text{and} \quad b_1{}^2 + b_2{}^2 + \cdots + b_n{}^2$$

from both sides, we see that this Pythagorean relation holds if and only if

$$a_1 b_1 + a_2 b_2 + \cdots + a_n b_n = 0.$$

This leads us at once to make the following definition. We wish to emphasize that the preceding computation was not to *prove* this definition (one doesn't prove a definition), but rather to *motivate* the definition.

Definition. Two vectors

$$\alpha = \langle a_1, a_2, \ldots, a_n \rangle \quad \text{and} \quad \beta = \langle b_1, b_2, \ldots, b_n \rangle$$

in $\mathbf{R}^n$ are **orthogonal** (**perpendicular**) if and only if

$$a_1 b_1 + a_2 b_2 + \cdots + a_n b_n = 0.$$

Example 3.5 The vectors $\langle 1, 0 \rangle$ and $\langle 0, 1 \rangle$ in $\mathbf{R}^2$ are orthogonal, for

$$1 \cdot 0 + 0 \cdot 1 = 0.$$

Also $\langle -1, 3, 2 \rangle$ and $\langle 5, -1, 4 \rangle$ are orthogonal vectors in $\mathbf{R}^3$, for

$$-1 \cdot 5 + 3 \cdot (-1) + 2 \cdot 4 = -5 - 3 + 8 = 0. \; \|$$

Example 3.6 According to our definition, the zero vector $\langle 0, 0, \ldots, 0 \rangle$ in $\mathbf{R}^n$ is orthogonal to *every* vector in $\mathbf{R}^n$. This is why it is convenient to think of $\langle 0, 0, \ldots, 0 \rangle$ as having *all directions*, rather than *no direction*. $\|$

EXERCISES

3.1 Consider the vectors $\alpha = \langle 2, -1 \rangle$ and $\beta = \langle -3, -2 \rangle$ in $\mathbf{R}^2$. Draw coordinate axes and sketch, using arrows, the vectors α, β, $\alpha + \beta$, $\alpha - \beta$, and $-\frac{4}{3}\alpha$.

3.2 Let $\alpha = \langle -1, 3, -2 \rangle$, $\beta = \langle 4, 0, -1 \rangle$ and $\gamma = \langle -3, -1, 2 \rangle$ be vectors in $\mathbf{R}^3$. Compute the following.

a) 3α b) -2γ c) $\alpha + \beta$ d) $3\beta - 2\gamma$
e) $\alpha + 2(\beta - 3\gamma)$ f) $3(\alpha - 2\beta)$ g) $4(3\alpha + 5\gamma)$

3.3 Consider the vectors $\alpha = \langle 3, -2, 2 \rangle$ and $\beta = \langle -1, 4, 1 \rangle$ in $\mathbf{R}^3$. Compute each of the following.

a) $\|\alpha\|$ b) $\|\alpha + \beta\|$ c) $\|-2\alpha\|$ d) $\|\beta - 3\alpha\|$

3.4 Determine whether the vectors in each of the following pairs are parallel, orthogonal, or neither. If parallel, state whether they have the same or opposite directions.

a) $\langle 3, -1 \rangle$ and $\langle 4, 12 \rangle$ in $\mathbf{R}^2$
b) $\langle -2, 6 \rangle$ and $\langle 4, -12 \rangle$ in $\mathbf{R}^2$
c) $\langle 3, -1, 0 \rangle$ and $\langle 4, 3, 2 \rangle$ in $\mathbf{R}^3$
d) $\langle 2, -3, 1 \rangle$ and $\langle 8, 2, -10 \rangle$ in $\mathbf{R}^3$
e) $\langle \sqrt{2}, \sqrt{18}, -\sqrt{8} \rangle$ and $\langle 2, 6, -4 \rangle$ in $\mathbf{R}^3$
f) $\langle -3, 8, 5, 1, 0 \rangle$ and $\langle 5, -1, 4, 3, 7 \rangle$ in $\mathbf{R}^5$
g) $\langle -1, 3, 4, -2, 1, 4 \rangle$ and $\langle 3, -4, -12, 6, -4, -2 \rangle$ in $\mathbf{R}^6$

3.5 If possible, determine x_2 so that the vector $\langle 2, x_2 \rangle$ in $\mathbf{R}^2$ is parallel to the given vector.

a) $\langle 4, 6 \rangle$ b) $\langle -5, 3 \rangle$ c) $\langle 3, 0 \rangle$ d) $\langle 0, 3 \rangle$

3.6 If possible, determine x_1 so that the vector $\langle x_1, 2, -1 \rangle$ is orthogonal to the given vector.

a) $\langle 0, 1, -4 \rangle$ b) $\langle 1, 0, -3 \rangle$ c) $\langle -5, 1, 2 \rangle$

3.7 Find a unit vector in $\mathbf{R}^4$ parallel to $\langle 1, 0, -1, 3 \rangle$ and having the same direction.

3.8 Find two unit vectors in $\mathbf{R}^2$ orthogonal to $\langle 3, -4 \rangle$.

3.9 Find two unit vectors in $\mathbf{R}^3$ which are not parallel and each of which is orthogonal to $\langle -2, 1, 2 \rangle$.

3.10 Show that $(1, -5)$, $(9, -11)$, and $(4, -1)$ are vertices of a right triangle in $\mathbf{R}^2$. (Use your high school idea of what is meant by a right triangle.)

3.11 Show that $(1, -1, 4)$, $(3, -2, 4)$, $(-4, 2, 6)$, and $(-2, 1, 6)$ are vertices of a parallelogram in $\mathbf{R}^3$. (Use your high school idea of what is meant by a parallelogram.)

3.12 Show that for all vectors $\alpha = \langle a_1, a_2, \ldots, a_n \rangle$, $\beta = \langle b_1, b_2, \ldots, b_n \rangle$, and $\gamma = \langle c_1, c_2, \ldots, c_n \rangle$ in $\mathbf{R}^n$, and all real numbers $r, s \in \mathbf{R}$, the following relations hold.

a) $(\alpha + \beta) + \gamma = \alpha + (\beta + \gamma)$ b) $\alpha + \beta = \beta + \alpha$
c) $r(s\alpha) = (rs)\alpha$ d) $(r + s)\alpha = r\alpha + s\alpha$
e) $r(\alpha + \beta) = r\alpha + r\beta$

4. LINES

4.1 Parametric Equations for a Line

You are probably accustomed to thinking of a line as being determined by two points. While this is perfectly correct, *it will be more useful for us to think of a line as being determined by one point on the line and the direction of the line.* Of course, the direction of a line can be specified in terms of a vector.

Figure 4.1 gives a picture of a line in $\mathbf{R}^2$ which passes through the origin and has a direction given by the direction vector $\prec d_1, d_2 \succ$. Clearly, if (x_1, x_2) is any point on this line, then the vector $\prec x_1, x_2 \succ$ equals $t \prec d_1, d_2 \succ$ for some $t \in \mathbf{R}$. Conversely, for each $t \in \mathbf{R}$, the point (td_1, td_2) is on the line.

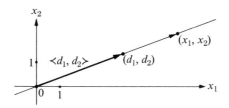

Fig. 4.1

In Fig. 4.2, we have illustrated for $\mathbf{R}^3$ a line passing through a point (a_1, a_2, a_3) and having a direction given by the direction vector

$$\delta = \prec d_1, d_2, d_3 \succ.$$

If $\alpha = \prec a_1, a_2, a_3 \succ$, we can think of $\alpha + t\delta$, for $t \in \mathbf{R}$, as a vector from the origin to a point on this line. Intuitively, α gets us from the origin to the point (a_1, a_2, a_3) on the line, and $t\delta$ then takes us along the line. For example, if $t = 0$, we have the vector α from the origin to (a_1, a_2, a_3). If $t = 1$, the vector $\alpha + \delta$ goes to the point $(a_1 + d_1, a_2 + d_2, a_3 + d_3)$, as illustrated in Fig. 4.2. Thus for a point (x_1, x_2, x_3) on this line, we must have

$$\prec x_1, x_2, x_3 \succ = \prec a_1, a_2, a_3 \succ + t \prec d_1, d_2, d_3 \succ$$
$$= \prec a_1 + d_1 t, a_2 + d_2 t, a_3 + d_3 t \succ$$

for some $t \in \mathbf{R}$, and, conversely, each $t \in \mathbf{R}$ gives rise to a point on the line.

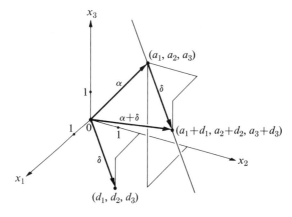

Fig. 4.2

The following definition generalizes these ideas to $\mathbf{R}^n$.

Definition. Let $(a_1, a_2, \ldots, a_n) \in \mathbf{R}^n$ and let

$$\delta = \prec d_1, d_2, \ldots, d_n \succ \neq \prec 0, 0, \ldots, 0 \succ$$

be a vector in $\mathbf{R}^n$. The ***line in $\mathbf{R}^n$ through*** $(a_1, a_2, \ldots, a_n)$ ***with direction*** δ is the set of all $(x_1, x_2, \ldots, x_n) \in \mathbf{R}^n$ such that for some $t \in \mathbf{R}$

$$\begin{aligned}
x_1 &= a_1 + d_1 t, \\
x_2 &= a_2 + d_2 t, \\
&\;\;\vdots \\
x_n &= a_n + d_n t.
\end{aligned} \tag{1}$$

Equations (1) are ***parametric equations for the line***, and t is the ***parameter***.

Example 4.1 Parametric equations for the line in $\mathbf{R}^3$ through $(-1, 0, 2)$ with direction $\prec 2, -3, -1 \succ$ are

$$\begin{aligned}
x_1 &= -1 + 2t, \\
x_2 &= -3t, \\
x_3 &= 2 - t. \;\|
\end{aligned}$$

Of course, if a line has direction δ, then it also has direction $r\delta$ for any nonzero $r \in \mathbf{R}$. *In general, if you want to find parametric equations of a line, you should always try to find a point on the line and a direction for the line, and use Eqs. (1).*

It is sometimes helpful to think of the parametric equations (1) as giving a rule for picking up each point t of a Euclidean 1-space (a t-axis, see Fig. 4.3) and putting it down in $\mathbf{R}^n$. The whole t-axis, that is, the whole Euclidean line, is picked up and put down in $\mathbf{R}^n$, with the origin 0 of $\mathbf{R}$ put down at $(a_1, a_2, \ldots, a_n) \in \mathbf{R}^n$. While Eqs. (1) do not "bend" the line as they pick it up and put it down in $\mathbf{R}^n$, they may "stretch" it if $\|\delta\| \neq 1$, for the point 1 is put down at $(a_1 + d_1, a_2 + d_2, \ldots, a_n + d_n)$ a distance of $\|\delta\|$ from $(a_1, a_2, \ldots, a_n)$.

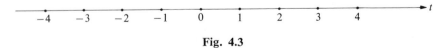

Fig. 4.3

The physicist usually thinks of the parameter t as representing *time*. Equations (1) can then be viewed as giving the location at time t of a body traveling along a line in $\mathbf{R}^n$. It is for this reason that one sometimes refers to $(a_1, a_2, \ldots, a_n)$ as the point on the line *when $t = 0$.*

4.2 Parallel and Orthogonal Lines

Definition. Two lines in $\mathbf{R}^n$ are *parallel* if they have parallel direction vectors, and are *orthogonal* if they have orthogonal direction vectors.

Of course, two lines in $\mathbf{R}^2$ either are parallel or intersect, but this need not be true in $\mathbf{R}^n$ for $n > 2$.

Example 4.2 The lines in $\mathbf{R}^2$ with equations

$$x_1 = -2 + t, \qquad x_2 = 3 - 2t$$

and

$$x_1 = 7 - 3s, \qquad x_2 = 1 + 2s$$

are not parallel, since the direction vector $\langle 1, -2 \rangle$ of the first line is not parallel to the direction vector $\langle -3, 2 \rangle$ of the second line. Let us find their point of intersection.

We used different parameters t and s for the two lines, for usually the point of intersection is not attained for the same values of the parameters on the two lines. If (a_1, a_2) lies on both lines, we must have

$$a_1 = -2 + t = 7 - 3s \quad \text{and} \quad a_2 = 3 - 2t = 1 + 2s$$

for some values of t and s. Therefore, we must solve simultaneously the equations

$$-2 + t = 7 - 3s \quad \text{and} \quad 3 - 2t = 1 + 2s$$

for t and s. These equations may be handled as follows. We have

$$t + 3s = 9, \qquad 2t + 2s = 2$$

or, dividing by the second equation by 2 and subtracting,

$$
\begin{aligned}
t + 3s &= 9 \\
t + s &= 1 \\
\hline
2s &= 8,
\end{aligned}
$$

so $s = 4$. Therefore $t = -3$. Thus the point of intersection can be found by either putting $s = 4$ or $t = -3$. In either case, the point $(-5, 9)$ is obtained. ‖

Example 4.3 Let us see whether the lines given by

$$x_1 = -2 + t, \qquad x_2 = 3 - 2t, \qquad x_3 = 1 + 5t$$

and

$$x_1 = 7 - 3s, \qquad x_2 = 1 + 2s, \qquad x_3 = 4 - s$$

in $\mathbf{R}^3$ intersect. In general, you would not expect two lines in $\mathbf{R}^3$ to intersect.

If (a_1, a_2, a_3) is a point on both lines, we would then have to have

$$a_1 = -2 + t = 7 - 3s,$$
$$a_2 = 3 - 2t = 1 + 2s,$$
$$a_3 = 1 + 5t = 4 - s.$$

Thus we try to find s and t such that

$$-2 + t = 7 - 3s, \quad 3 - 2t = 1 + 2s, \quad \text{and} \quad 1 + 5t = 4 - s.$$

Since there are only two quantities, t and s, to find, we select two equations and try to find t and s. Take for example the equations

$$-2 + t = 7 - 3s \quad \text{and} \quad 3 - 2t = 1 + 2s.$$

Then Example 4.2 shows that we must have $t = -3$ and $s = 4$. But $t = -3$ gives the point $(-5, 9, -14)$, while $s = 4$ gives $(-5, 9, 0)$. Thus the lines do not intersect. ‖

If $(a_1, a_2, \ldots, a_n)$ and $(b_1, b_2, \ldots, b_n)$ are two distinct points in $\mathbf{R}^n$, then clearly

$$\delta = \langle b_1 - a_1, b_2 - a_2, \ldots, b_n - a_n \rangle$$

is a direction vector for the line joining these points. Therefore

$$x_1 = a_1 + (b_1 - a_1)t,$$
$$x_2 = a_2 + (b_2 - a_2)t,$$
$$\cdot$$
$$\cdot \qquad\qquad (2)$$
$$\cdot$$
$$x_n = a_n + (b_n - a_n)t$$

are parametric equations of a line which passes through $(a_1, a_2, \ldots, a_n)$, when $t = 0$, and also through $(b_1, b_2, \ldots, b_n)$, when $t = 1$. This shows how to find parametric equations for a line given by two points. You should think of Eqs. (2) as being determined by the point $(a_1, a_2, \ldots, a_n)$ on the line and the vector $\delta = \langle b_1 - a_1, b_2 - a_2, \ldots, b_n - a_n \rangle$ which has a direction and length given by an arrow starting at $(a_1, a_2, \ldots, a_n)$ and having its tip at $(b_1, b_2, \ldots, b_n)$.

Example 4.4 Parametric equations for the line in $\mathbf{R}^4$ determined by the points $(2, -1, 3, 6)$ and $(-1, 0, 3, -4)$ are

$$x_1 = 2 - 3t, \quad x_2 = -1 + t, \quad x_3 = 3, \quad x_4 = 6 - 10t$$

or, equivalently,

$$x_1 = -1 + 3t, \quad x_2 = -t, \quad x_3 = 3, \quad x_4 = -4 + 10t. \ ‖$$

*4.3 Line Segments and Convex Sets

Finally, we describe analytically the *line segment* between two points $(a_1, a_2, \ldots, a_n)$ and $(b_1, b_2, \ldots, b_n)$ in $\mathbf{R}^n$. Let $\alpha = \langle a_1, a_2, \ldots, a_n \rangle$ and $\beta = \langle b_1, b_2, \ldots, b_n \rangle$. We can think of the direction and magnitude of the vector $\beta - \alpha = \langle b_1 - a_1, b_2 - a_2, \ldots, b_n - a_n \rangle$ as being represented by an arrow from $(a_1, a_2, \ldots, a_n)$ to $(b_1, b_2, \ldots, b_n)$. Thus, as illustrated in Fig. 4.4, we can reach the tip of $\alpha + t(\beta - \alpha)$ by going to the tip of α and then moving in the direction of $\beta - \alpha$ for the distance given by $t(\beta - \alpha)$. The values of t between 0 and 1 correspond to points which we consider to be on the line segment; for example, $t = 0$ gives $(a_1, a_2, \ldots, a_n)$ and $t = 1$ gives $(b_1, b_2, \ldots, b_n)$. If $(x_1, x_2, \ldots, x_n)$ is on the line segment, we would have $x_i = a_i + (b_i - a_i)t$ for some value of t between 0 and 1. In particular, $t = \frac{1}{2}$ gives the point halfway between $(a_1, a_2, \ldots, a_n)$ and $(b_1, b_2, \ldots, b_n)$.

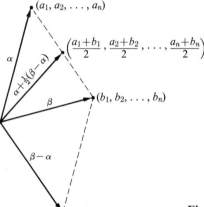

Fig. 4.4

Definition. Let $(a_1, a_2, \ldots, a_n)$ and $(b_1, b_2, \ldots, b_n)$ be two points in $\mathbf{R}^n$. The *line segment joining them* consists of all points $(x_1, x_2, \ldots, x_n) \in \mathbf{R}^n$ such that for some values of t, where $0 \leq t \leq 1$,

$$x_1 = a_1 + (b_1 - a_1)t,$$
$$x_2 = a_2 + (b_2 - a_2)t,$$
$$\vdots$$
$$x_n = a_n + (b_n - a_n)t.$$

The *midpoint of the line segment* is the point

$$\left(\frac{a_1 + b_1}{2}, \frac{a_2 + b_2}{2}, \ldots, \frac{a_n + b_n}{2} \right).$$

A subset S of $\mathbf{R}^n$ is **convex** if for any two points in S, the entire line segment in $\mathbf{R}^n$ joining the points is again in S.

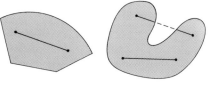

Convex subset   Nonconvex subset   **Fig. 4.5**

Figure 4.5 gives an illustration of a convex subset and a nonconvex subset of $\mathbf{R}^2$.

EXERCISES

4.1 Give parametric equations for the line in $\mathbf{R}^2$ through $(3, -2)$ with direction $\prec -8, 4 \succ$. Sketch the line in an appropriate figure.

4.2 Give parametric equations for the line in $\mathbf{R}^3$ through $(-1, 3, 0)$ with direction $\prec -2, -1, 4 \succ$. Sketch the line in an appropriate figure.

4.3 Give parametric equations for the line in $\mathbf{R}^4$ through $(2, -1, 0, 3)$ with direction $\prec -1, 0, 4, -3 \succ$.

4.4 Consider the line in $\mathbf{R}^2$ with parametric equations

$$x_1 = -3 + 4t \qquad \text{and} \qquad x_2 = 2 - 3t.$$

a) Find a direction vector for the line.
b) Find the point on the line whose x_1-coordinate is 1.
c) Find the point on the line whose x_2-coordinate is 8.

4.5 Find parametric equations for the line in $\mathbf{R}^2$ through $(5, -1)$ and orthogonal to the line with parametric equations $x_1 = 4 - 2t$ and $x_2 = 7 + t$.

4.6 For each pair of points, find parametric equations of the line containing them.
a) $(-2, 4)$ and $(3, -1)$ in $\mathbf{R}^2$ b) $(3, -1, 6)$ and $(0, -3, -1)$ in $\mathbf{R}^3$
c) $(0, 0, 0, 0)$ and $(4, 1, -1, 3)$ in $\mathbf{R}^4$

4.7 For each pair of lines in $\mathbf{R}^2$, determine whether the lines are parallel or intersect. If they intersect, find the point of intersection, and determine whether the lines are orthogonal.

a) $\begin{cases} x_1 = 5 + 3t, \\ x_2 = -6 - 5t \end{cases}$ and $\begin{cases} x_1 = -5 + 4s, \\ x_2 = 5 - s \end{cases}$

b) $\begin{cases} x_1 = 6 - 2t, \\ x_2 = -4 + 4t \end{cases}$ and $\begin{cases} x_1 = -3 + 3s, \\ x_2 = 7 - 6s \end{cases}$

c) $\begin{cases} x_1 = -3 + 3t, \\ x_2 = -11 + 9t \end{cases}$ and $\begin{cases} x_1 = 12 - 6s, \\ x_2 = -6 + 2s \end{cases}$

4.8 For each of the given pairs of lines in $\mathbf{R}^3$, determine whether the lines intersect. If they intersect, find the point of intersection, and determine whether the lines are orthogonal.

a) $\begin{cases} x_1 = & 4 + t, \\ x_2 = & 2 - 3t, \\ x_3 = & -3 + 5t \end{cases}$ and $\begin{cases} x_1 = & 11 + 3s, \\ x_2 = & -9 - 4s, \\ x_3 = & -4 - 3s \end{cases}$

b) $\begin{cases} x_1 = & 11 + 3t, \\ x_2 = & -3 - t, \\ x_3 = & 4 + 3t \end{cases}$ and $\begin{cases} x_1 = & 6 - 2s, \\ x_2 = & -2 + s, \\ x_3 = & -15 + 7s \end{cases}$

4.9 Find all points in common to the lines in $\mathbf{R}^2$ given by

$$x_1 = 5 - 3t, \qquad x_2 = -1 + t$$

and

$$x_1 = -7 + 6s, \qquad x_2 = 3 - 2s.$$

4.10 Find parametric equations for the line in $\mathbf{R}^3$ through $(-1, 2, 3)$ which is orthogonal to each of the two lines having parametric equations

$$x_1 = -2 + 3t, \qquad x_2 = 4, \qquad x_3 = 1 - t$$

and

$$x_1 = 7 - t, \qquad x_2 = 2 + 3t, \qquad x_3 = 4 + t.$$

*4.11 Find the midpoint of the line segment joining each of the following pairs of points.

a) $(-2, 4)$ and $(3, -1)$ in $\mathbf{R}^2$ b) $(3, -1, 6)$ and $(0, -3, -1)$ in $\mathbf{R}^3$

c) $(0, 0, 0, 0)$ and $(4, 1, -1, 3)$ in $\mathbf{R}^4$

*4.12 Find the point in $\mathbf{R}^2$ on the line segment joining $(-1, 3)$ and $(2, 5)$ which is twice as close to $(-1, 3)$ as to $(2, 5)$.

*4.13 Find the point in $\mathbf{R}^4$ on the line segment joining $(-2, 1, 3, -2)$ and $(0, -5, 2, 6)$ which is one-fourth of the way from $(-2, 1, 3, -2)$ to $(0, -5, 2, 6)$.

*4.14 Sketch the smallest convex subset of $\mathbf{R}^2$ containing the given points.

a) $(-1, 3)$ and $(4, -1)$ b) $(-1, 3)$, $(4, -1)$, and $(2, 5)$
c) $(0, 1)$, $(1, 0)$, $(0, -1)$, and $(-1, 0)$ d) $(0, 0)$, $(1, 1)$, $(-1, 1)$, and $(0, -1)$

*4.15 Let $(a_1, a_1, \ldots, a_n)$ and $(b_1, b_2, \ldots, b_n)$ be any two points in $\mathbf{R}^n$. Show that the line segment joining the points consists of all points $(x_1, x_2, \ldots, x_n)$, where $x_i = (1 - t)a_i + tb_i$ for $0 \le t \le 1$.

5. HYPERPLANES

5.1 Definition of a Hyperplane in $\mathbf{R}^n$

The line in $\mathbf{R}^n$ with parametric equations

$$x_1 = a_1 + d_1 t,$$
$$x_2 = a_2 + d_2 t,$$
$$\cdot$$
$$\cdot$$
$$\cdot$$
$$x_n = a_n + d_n t$$

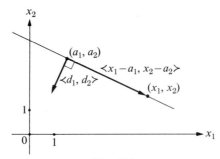

Fig. 5.1

can be characterized as the set of all points $(x_1, x_2, \ldots, x_n) \in \mathbf{R}^n$ such that the vector $\langle x_1 - a_1, x_2 - a_2, \ldots, x_n - a_n \rangle$ is *parallel* to the direction vector $\delta = \langle d_1, d_2, \ldots, d_n \rangle$ of the line (see Fig. 5.1). Let us consider now the set of all $(x_1, x_2, \ldots, x_n) \in \mathbf{R}^n$ such that the vector

$$\langle x_1 - a_1, x_2 - a_2, \ldots, x_n - a_n \rangle$$

is *orthogonal* to the vector $\delta = \langle d_1, d_2, \ldots, d_n \rangle$, where not all d_i are zero. By definition, these two vectors are orthogonal if and only if

$$d_1(x_1 - a_1) + d_2(x_2 - a_2) + \cdots + d_n(x_n - a_n) = 0. \tag{1}$$

We should try to see in $\mathbf{R}^2$ and $\mathbf{R}^3$ whether this concept gives us some familiar geometric configuration.

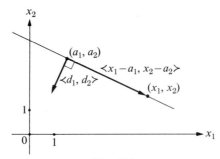

Fig. 5.2

From Fig. 5.2, we see that all such points $(x_1, x_2) \in \mathbf{R}^2$ lie on the line through (a_1, a_2) which is perpendicular to δ, and from Fig. 5.3, we see that all such points $(x_1, x_2, x_3) \in \mathbf{R}^3$ lie in a plane through (a_1, a_2, a_3). It is harder to see what happens in $\mathbf{R}^n$, but if we take the special case

$$(a_1, a_2, \ldots, a_n) = (0, 0, \ldots, 0)$$

and

$$\langle d_1, d_2, \ldots, d_n \rangle = \langle 0, 0, \ldots, 0, 1 \rangle,$$

Eq. (1) reduces to

$$1(x_n - 0) = 0 \qquad \text{or} \qquad x_n = 0.$$

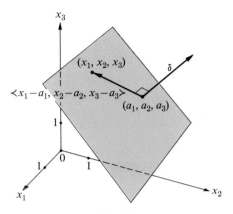

Fig. 5.3

Clearly, $\{(x_1, x_2, \ldots, x_{n-1}, 0) \mid x_i \in \mathbf{R}\}$ is essentially an $(n-1)$-dimensional Euclidean space, for the 0 in the final position is just carried along with the right-hand parenthesis. These examples illustrate that the set of all solutions of Eq. (1) always gives a line in $\mathbf{R}^2$ for $n = 2$, a plane in $\mathbf{R}^3$ for $n = 3$, and, in general, an $(n-1)$-dimensional "flat piece" of $\mathbf{R}^n$.

> **Definition.** Let $(a_1, a_2, \ldots, a_n) \in \mathbf{R}^n$ and let $\delta = \prec d_1, d_2, \ldots, d_n \succ$, where not all d_i are zero. The set of all $(x_1, x_2, \ldots, x_n) \in \mathbf{R}^n$ such that $d_1(x_1 - a_1) + d_2(x_2 - a_2) + \cdots + d_n(x_n - a_n) = 0$ is the **hyperplane in $\mathbf{R}^n$ passing through $(a_1, a_2, \ldots, a_n)$ and having the orthogonal direction** δ. The defining equation is an **equation of the hyperplane.**

One visualizes the direction of a hyperplane as being perpendicular to the hyperplane.

You should note the formal similarity of this definition of a hyperplane with the definition of a line in the last section. A line is a one-dimensional "flat piece" of $\mathbf{R}^n$, while a hyperplane is an $(n-1)$-dimensional "flat piece" of $\mathbf{R}^n$. Without any further explanation, we tell you that, in general, for $\mathbf{R}^n$ you can expect the definition of an r-dimensional "flat piece" to be quite similar to the definition of an $(n-r)$-dimensional "flat piece."

Example 5.1 Let us find an equation of the plane in $\mathbf{R}^3$ which passes through $(-1, 2, 1)$ and has the (orthogonal) direction vector $\prec 1, -3, 2 \succ$. Equation (1) becomes

$$1(x_1 - (-1)) + (-3)(x_2 - 2) + 2(x_3 - 1) = 0,$$

or

$$x_1 - 3x_2 + 2x_3 = -5. \; \|$$

Of course if δ is a direction vector for a hyperplane, then $r\delta$ is also a direction vector for the same hyperplane for all $r \in \mathbf{R}$.

Just as in Example 5.1, Eq. (1) can always be rewritten in the form

$$d_1x_1 + d_2x_2 + \cdots + d_nx_n = c,$$

where $c = d_1a_1 + d_2a_2 + \cdots + d_na_n$ is in $\mathbf{R}$. The following theorem gives the converse of this observation.

Theorem 5.1 *If $d_1, d_2, \ldots, d_n$ are real numbers not all zero, then for any $c \in \mathbf{R}$, the set of all points $(x_1, x_2, \ldots, x_n) \in \mathbf{R}^n$ such that*

$$d_1x_1 + d_2x_2 + \cdots + d_nx_n = c$$

is a hyperplane in $\mathbf{R}^n$ with (orthogonal) direction vector $\prec d_1, d_2, \ldots, d_n \succ$.

Proof. By assumption, not all d_i are zero; suppose $d_1 \neq 0$. Choose any numbers $a_2, a_3, \ldots, a_n \in \mathbf{R}$, and let

$$a_1 = \frac{c - d_2a_2 - \cdots - d_na_n}{d_1}.$$

Then $d_1a_1 + d_2a_2 + \cdots + d_na_n = c$. Now for any $(x_1, x_2, \ldots, x_n) \in \mathbf{R}^n$ such that $d_1x_1 + d_2x_2 + \cdots + d_nx_n = c$, we have the two equations

$$d_1x_1 + d_2x_2 + \cdots + d_nx_n = c,$$

$$d_1a_1 + d_2a_2 + \cdots + d_na_n = c.$$

Subtracting, we get

$$d_1(x_1 - a_1) + d_2(x_2 - a_2) + \cdots + d_n(x_n - a_n) = 0.$$

Thus $(x_1, x_2, \ldots, x_n)$ lies on the hyperplane through $(a_1, a_2, \ldots, a_n)$ with direction vector $\prec d_1, d_2, \ldots, d_n \succ$. ∎

Example 5.2 By Theorem 5.1, the set of all $(x_1, x_2) \in \mathbf{R}^2$ such that

$$0x_1 + 1x_2 = 0$$

should be a hyperplane in $\mathbf{R}^2$, that is, a line. This line contains $(0, 0)$ and has as (orthogonal) direction vector $\prec 0, 1 \succ$. Thus this line is simply the x_1-axis. Of course, it is clear that the solutions in $\mathbf{R}^2$ of $x_2 = 0$ are precisely the points $(a, 0) \in \mathbf{R}^2$. ‖

Example 5.3 As in Example 5.2, it is easy to see that the equation $x_3 = 0$ gives in $\mathbf{R}^3$ the plane through the origin which contains the x_1- and x_2-coordinate axes. ‖

You can easily check that for $\mathbf{R}^2$ and $\mathbf{R}^3$ the following definitions coincide with your intuitive ideas of parallel and perpendicular (orthogonal). One may remember these definitions by visualizing the situation in $\mathbf{R}^3$.

Definition. Two hyperplanes in $\mathbf{R}^n$ are *parallel* if direction vectors for the two hyperplanes are parallel. The hyperplanes are *orthogonal* if their direction vectors are orthogonal. A line and a hyperplane in $\mathbf{R}^n$ are *orthogonal* if the line has a direction vector which is also a direction vector of the hyperplane.

5.2 Computations

The following examples illustrate a few of the many types of problems we can now solve. Remember:

> *To find parametric equations for a line, you want to know a point on the line and a direction vector of the line.*
>
> *To find an equation of a hyperplane, you want to know a point in the hyperplane and an (orthogonal) direction vector for the hyperplane.*

Example 5.4 Let us find parametric equations for the line in $\mathbf{R}^3$ passing through the point $(1, 2, -1)$ and orthogonal to the plane with equation

$$3x_1 + 5x_2 - x_3 = 6.$$

We know a point $(1, 2, -1)$ on the line, and we need a direction vector for the line. Since the line is to be orthogonal to the plane with equation

$$3x_1 + 5x_1 - x_3 = 6,$$

we see by Theorem 5.1 that $\langle 3, 5, -1 \rangle$ is a direction vector for the line. Thus the line has parametric equations

$$x_1 = 1 + 3t, \qquad x_2 = 2 + 5t, \qquad x_3 = -1 - t. \parallel$$

Example 5.5 Let us find all points of intersection in $\mathbf{R}^3$ of the plane with equation

$$3x_1 + 5x_2 - x_3 = -2$$

and the line with parametric equations

$$x_1 = -3 + 2t, \qquad x_2 = 4 + t, \qquad x_3 = -1 - 3t.$$

If (x_1, x_2, x_3) lies on both the line and the plane, then substituting, we must have

$$3(-3 + 2t) + 5(4 + t) - (-1 - 3t) = -2$$

so

$$14t = -14$$

and

$$t = -1.$$

Thus the only point of intersection is $(-5, 3, 2)$. $\parallel$

Example 5.6 Let us find the distance in $\mathbf{R}^4$ from the point $(-9, 6, 2, 7)$ to the hyperplane $5x_1 - 2x_2 + x_3 - 3x_4 = 2$.

We have not defined what is meant by the distance from a point to a hyperplane. As you might guess from $\mathbf{R}^2$ and $\mathbf{R}^3$, the shortest distance from a point to a hyperplane is measured along the line which passes through the point and is orthogonal to the hyperplane. In our case, since $\langle 5, -2, 1, -3 \rangle$ is an (orthogonal) direction vector for the hyperplane, we find that parametric equations of this line are

$$x_1 = -9 + 5t, \qquad x_2 = 6 - 2t, \qquad x_3 = 2 + t, \qquad x_4 = 7 - 3t.$$

Let us find the point of intersection of this line with the given hyperplane. We have

$$5(-9 + 5t) - 2(6 - 2t) + (2 + t) - 3(7 - 3t) = 2$$

or

$$39t = 78.$$

Thus $t = 2$, and the line intersects the hyperplane at $(1, 2, 4, 1)$. For our answer, we therefore find the distance from $(-9, 6, 2, 7)$ to $(1, 2, 4, 1)$, namely

$$\sqrt{(10)^2 + 4^2 + 2^2 + 6^2} = \sqrt{156}. \ \|$$

EXERCISES

5.1 a) What constitutes a line in $\mathbf{R} = \mathbf{R}^1$, according to our definition?
b) What constitutes a hyperplane in $\mathbf{R} = \mathbf{R}^1$, according to our definition?

5.2 Find an equation for the hyperplane in $\mathbf{R}^4$ through $(1, -3, 0, 2)$ and with direction vector $\langle 0, -1, 4, 3 \rangle$.

5.3 Find an equation of the plane in $\mathbf{R}^3$ parallel to the plane with equation $3x_1 - 2x_2 + 7x_3 = 14$ and passing through $(3, -1, 4)$.

5.4 Find an equation of the hyperplane in $\mathbf{R}^4$ passing through $(-1, 4, -3, 2)$ and orthogonal to the line with parametric equations $x_1 = 3 - 7t$, $x_2 = 4 + t$, $x_3 = 2t$, $x_4 = 2 - t$.

5.5 Find an equation of the plane in $\mathbf{R}^3$ passing through the origin and orthogonal to the line through the points $(-1, 3, 0)$ and $(2, -4, 3)$.

5.6 Classify the hyperplanes given by the following pairs of equations as parallel, orthogonal, or neither.

a) $x_1 + 4x_2 - 2x_4 = 7$ and $x_2 - x_3 + 2x_4 = -3$ in $\mathbf{R}^4$
b) $3x_1 + 4x_2 - x_3 = 1$ and $-6x_1 - 8x_2 + 2x_3 = 4$ in $\mathbf{R}^3$
c) $x_1 - 3x_2 = 7$ and $2x_1 + 4x_2 = 1$ in $\mathbf{R}^2$
d) $x_1 - 3x_3 = 8$ and $x_2 + 4x_4 + x_5 = 2$ in $\mathbf{R}^6$

5.7 Find parametric equations of the line in $\mathbf{R}^4$ which is orthogonal to the hyperplane with equation $x_1 - 2x_2 + x_3 + 4x_4 = 3$ and which passes through the point $(-2, 1, 0, 5)$.

5.8 Find the intersection in $\mathbf{R}^3$ of the line given by $x_1 = 5 + t$, $x_2 = -3t$, $x_3 = -2 + 4t$ and the plane with equation $x_1 - 3x_2 + 2x_3 = -25$.

5.9 Find the intersection in $\mathbf{R}^5$ of the line given by $x_1 = 2$, $x_2 = 5 - t$, $x_3 = 2t$, $x_4 = -t$, $x_5 = 2 + t$ and the hyperplane with equation $x_1 + 2x_3 - x_5 = 6$.

5.10 Find the distance in $\mathbf{R}^3$ from the point $(1, 3, -1)$ to the plane with equation $2x_1 - x_2 + x_3 = 4$.

5.11 Describe the geometric significance of each of the following conditions on the coefficients of the x_i in the equation of a hyperplane.

a) The coefficient of x_1 is zero for the equation of a hyperplane in $\mathbf{R}^2$.
b) The coefficient of x_1 is zero for the equation of a hyperplane in $\mathbf{R}^3$.
c) The coefficient of x_1 is zero for the equation of a hyperplane in $\mathbf{R}^4$.

d) The coefficients of both x_1 and x_2 are zero for the equation of a hyperplane in $\mathbf{R}^3$.

e) The coefficients of both x_1 and x_2 are zero for the equation of a hyperplane in $\mathbf{R}^4$.

5.12 Find an equation of the plane which passes through $(1, 0, 0)$, $(0, 1, 0)$, and $(0, 0, 1)$ in $\mathbf{R}^3$. [*Hint:* Determine a, b, c, and d so that these points satisfy the equation $ax_1 + bx_2 + cx_3 = d$.]

5.13 Find an equation of the hyperplane in $\mathbf{R}^4$ which passes through $(1, 0, 0, 0)$, $(0, 1, 0, 0)$, $(0, 0, 1, 0)$, and $(0, 0, 0, 1)$. [*Hint:* Proceed as in the hint for Exercise 5.12.]

5.14 Find an equation of the plane in $\mathbf{R}^3$ which passes through $(1, 0, 0)$, $(0, 1, -1)$, and $(1, 1, 1)$. [*Hint:* Proceed as in the hint for Exercise 5.12.]

6. PEDAGOGICAL CONSIDERATIONS: SLOPES

6.1 Some Pedagogical Considerations

We cannot resist the temptation to give some comments on pedagogy in the final section of this chapter. A line in $\mathbf{R}^n$ is a one-dimensional "flat piece" of $\mathbf{R}^n$, and a hyperplane is an $(n - 1)$-dimensional "flat piece." The concepts of a line and a hyperplane coincide in $\mathbf{R}^2$, since $1 = 2 - 1$. Most texts on analytic geometry start with the study of $\mathbf{R}^2$ and *define* a line in $\mathbf{R}^2$ to be a hyperplane, that is, to be the set of all $(x_1, x_2) \in \mathbf{R}^2$ such that for certain a, b, $c \in \mathbf{R}$, where a and b are not both zero,

$$ax_1 + bx_2 = c.$$

The student is then drilled that a line is the set of solutions of such a single linear equation. Thus when the student studies $\mathbf{R}^3$ later, he expects a line in $\mathbf{R}^3$ to be given by a single linear equation $ax_1 + bx_2 + cx_3 = d$, *and of course this is wrong*!

A line in $\mathbf{R}^3$ can be defined as the intersection of two planes which are not parallel, so a line in $\mathbf{R}^3$ can be described as the set of common solutions of *two* linear equations in x_1, x_2, and x_3, provided the coefficients of x_1, x_2, and x_3 in the two equations are not proportional, so that the planes are not parallel. The advantage of our parametric approach to a line in $\mathbf{R}^2$ lies in the ease with which the definition is generalized to higher dimensions. One simply throws in one new equation for each dimension.

It is a fact that any "flat $(n - r)$-dimensional piece" of $\mathbf{R}^n$ can be obtained by an intersection of r hyperplanes, but you have to be careful that the hyperplanes come together just right; for example, no two can be parallel. In particular, taking $r = n - 1$, a line in $\mathbf{R}^n$ can be defined as an intersection of $n - 1$ suitably selected hyperplanes of $\mathbf{R}^n$. Also, an r-dimensional "flat piece" of $\mathbf{R}^n$ can be obtained parametrically in terms of r parameters $t_1, t_2, \ldots, t_r$, but again one has to be a bit careful to be sure that the dimension of the "flat piece" obtained is r, and not less than r. For $r = 1$, for example, we required that not all d_i were zero in the direction vector $\langle d_1, d_2, \ldots, d_n \rangle$. Corresponding conditions for $r > 1$ are more complicated.

From the pedagogical viewpoint, there is no question but that a line in $\mathbf{R}^n$, a one-dimensional "flat piece," is best defined parametrically, and a hyperplane, an $(n-1)$-dimensional "flat piece" of $\mathbf{R}^n$, is best defined as the set of solutions of a single linear equation.

6.2 Slopes

Almost every treatment of plane analytic geometry emphasizes the concept of the *slope m of a line*. Recall that a line is determined by a point on the line and a direction vector $\delta = \langle d_1, d_2 \rangle$ for the line. The slope of a line essentially determines its direction and represents an attempt to make one number m do the work of the two numbers d_1 and d_2. This attempt fails, since some lines have no slope. Also, slopes are unsatisfactory because they do not generalize to Euclidean spaces of dimension greater than 2. We present the notion of the slope of a line in $\mathbf{R}^2$ only so that you can refer more easily to other books.

If $\delta = \langle d_1, d_2 \rangle$ is a direction vector for a line, d_1 and d_2 cannot both be zero. If $d_1 \neq 0$, then

$$\frac{1}{d_1} \langle d_1, d_2 \rangle = \left\langle 1, \frac{d_2}{d_1} \right\rangle$$

is also a direction vector for the line. Of course, $d_1 = 0$ if and only if the line is vertical in our usual picture of $\mathbf{R}^2$. The number d_2/d_1 is the *slope of the line*.

Definition. If a line in $\mathbf{R}^2$ has direction vector $\langle d_1, d_2 \rangle$, where $d_1 \neq 0$, then the number $m = d_2/d_1$ is the **slope of the line**. A (vertical) line with direction vector $\langle 0, d_2 \rangle$ has **no slope**.

As shown in Fig. 6.1, the slope m of a line which is not vertical is the number of units the line rises if one step is taken horizontally in the positive

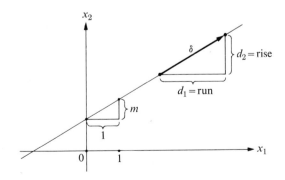

Fig. 6.1

x_1-direction. Some texts define the slope to be the quotient

$$\frac{rise}{run}$$

(see Fig. 6.1), where *rise* is measured vertically and *run* is measured horizontally. The term *slope* is a natural one to use; it has a meaning close to its meaning in everyday usage.

   If a line rises as you go to the right, then $m > 0$, while if the line falls, $m < 0$ (see Fig. 6.2). If $m = 0$, then the line is horizontal.

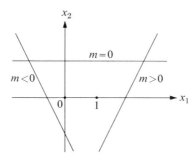

Fig. 6.2

Example 6.1 Let us find the slope of the line in $\mathbf{R}^2$ with parametric equations $x_1 = 2 - 3t$, $x_2 = 4 + 6t$.

   A direction vector δ for the line is $\langle -3, 6 \rangle$. Thus the slope m is $6/-3 = -2$. $\parallel$

   If a line in $\mathbf{R}^2$ passes through the points (a_1, a_2) and (b_1, b_2), then $\delta = \langle b_1 - a_1, b_2 - a_2 \rangle$ is a direction vector for the line. The slope of this line is therefore

$$m = \frac{b_2 - a_2}{b_1 - a_1},$$

provided that $b_1 - a_1 \neq 0$.

Example 6.2 The slope of the line through the points $(-3, 2)$ and $(0, 4)$ is

$$m = \frac{4 - 2}{0 - (-3)} = \frac{2}{3}. \parallel$$

   If a line in $\mathbf{R}^2$ has direction vector $\delta = \langle d_1, d_2 \rangle$, then $\langle d_2, -d_1 \rangle$ is a direction vector orthogonal to δ, since

$$d_1 d_2 + d_2(-d_1) = 0.$$

If the line passes through (a_1, a_2), we may view it as a hyperplane through (a_1, a_2) with (orthogonal) direction $\langle d_2, -d_1 \rangle$, and an equation for this hyperplane (line) is then

$$d_2(x_1 - a_1) - d_1(x_2 - a_2) = 0 \quad \text{or} \quad d_1 x_2 = d_2 x_1 + (d_1 a_2 - d_2 a_1).$$

If $d_1 \neq 0$, we may write this equation as

$$x_2 = \frac{d_2}{d_1} x_1 + \frac{d_1 a_2 - d_2 a_1}{d_1} .$$

Noting that $m = d_2/d_1$ and setting $b = (d_1 a_2 - d_2 a_1)/d_1$, we obtain the equation

$$x_2 = mx_1 + b. \tag{1}$$

The point $(0, b)$ obviously satisfies this equation, so the line (hyperplane) crosses the x_2-axis at b; that is, b is the x_2-*intercept of the line*. Actually, the x_1- and x_2-axes are classically termed the "x- and y-axes," respectively, so the classical form for Eq. (1) is

$$y = mx + b. \tag{2}$$

Equation (2) is classically known as the **slope-intercept equation of the line**. The number b is the **y-intercept of the line**.

Example 6.3 Let us find the classical slope-intercept equation [Eq. (2)] for the line in $\mathbf{R}^2$ which passes through the points $(-1, 4)$ and $(2, -3)$.

The slope of this line is

$$m = \frac{(-3) - 4}{2 - (-1)} = \frac{-7}{3} .$$

Thus Eq. (2) takes the form

$$y = -\tfrac{7}{3}x + b,$$

and we only need to find b. The value of b may be determined by the condition that $(-1, 4)$ satisfies the equation. Substituting, we obtain

$$4 = -\tfrac{7}{3}(-1) + b$$

so

$$b = 4 - \tfrac{7}{3} = \tfrac{12}{3} - \tfrac{7}{3} = \tfrac{5}{3}.$$

Thus the classical slope-intercept equation is

$$y = -\tfrac{7}{3}x + \tfrac{5}{3}. \parallel$$

Example 6.4 Let us find a vector orthogonal to the line given in Example 6.3. From the slope-intercept form which we found, we obtain the equation

$$3y = -7x + 5 \quad \text{or} \quad 7x + 3y = 5,$$

which we may view as an equation of a hyperplane in $\mathbf{R}^2$. By Section 5, a vector orthogonal to this hyperplane is $\langle 7, 3 \rangle$. $\parallel$

Suppose two lines have slopes m_1 and m_2. Then $\langle 1, m_1 \rangle$ and $\langle 1, m_2 \rangle$ are direction vectors for the lines, so the lines are orthogonal if and only if

$$1 + m_1 m_2 = 0,$$

or if and only if

$$m_2 = -\frac{1}{m_1}.$$

Of course, two nonvertical lines are parallel if and only if they have the same slope.

Example 6.5 Any line orthogonal to the line given in Example 6.3 has slope

$$-\frac{1}{-\frac{7}{3}} = \frac{3}{7},$$

while any line parallel to the line in Example 6.3 has slope $-\frac{7}{3}$. ‖

We wish to emphasize that the *proper* way to think of a line in $\mathbf{R}^2$, as well as in $\mathbf{R}^3$, $\mathbf{R}^4$, etc., is *parametrically*, as we explained in Section 4. Don't become enamored with these classical slopes in $\mathbf{R}^2$. We presented this $\mathbf{R}^2$-concept only so that you would understand its meaning if you run into it elsewhere.

EXERCISES

6.1 Find equations of two planes in $\mathbf{R}^3$ whose intersection is the line given by $x_1 = 6 - 3t, x_2 = 4 + t, x_3 = -1 + 2t$. [*Hint:* Eliminate t from the equations.]

6.2 Find equations of three hyperplanes in $\mathbf{R}^4$ whose intersection is the line given by $x_1 = 2$, $x_2 = -4 + t$, $x_3 = 1 - 2t$, $x_4 = -5$. [*Hint:* Eliminate t from the equations.]

6.3 Find a single linear equation $ax_1 + bx_2 = c$ which has as solution set in $\mathbf{R}^2$ the line through $(-1, 4)$ parallel to the line given by $x_1 = -2 + 3t, x_2 = 4 - t$. [*Hint:* Find parametric equations of the line, and eliminate the parameter t from the equations.]

6.4 For each of the given lines in $\mathbf{R}^2$, find its slope, if it exists.

a) The line through $(1, 1)$ with direction vector $<-3, 4>$
b) The line through $(-1, 0)$ with direction vector $<3, 0>$
c) The line through $(2, -1)$ with direction vector $<0, 3>$
d) The line through $(2, 4)$ and $(5, 7)$
e) The line through $(-2, 4)$ and $(3, -4)$
f) The line through $(3, -4)$ and $(3, 7)$
g) The line with parametric equations $x_1 = 3 - 7t, x_2 = 4 + 2t$
h) The line with parametric equations $x_1 = 4, x_2 = 4 + 2t$
i) The line with parametric equations $x_1 = 4 + t, x_2 = 6$

6.5 Find parametric equations of the line through $(-2, 4)$ with slope -2.

6.6 Find the classical slope-intercept equation [Eq. (2)] of the line through $(-2, 4)$ with slope -2.

6.7 Find the classical slope-intercept equation [Eq. (2)] of the line through $(1, -3)$ and $(2, 7)$.

6.8 Find the classical slope-intercept equation [Eq. (2)] of the line with parametric equations $x_1 = 3 + 4t, x_2 = -1 + 2t$.

6.9 Find parametric equations of the line with classical slope-intercept equation $y = 2x - 4$.

6.10 Find the classical slope-intercept equation [Eq. (2)] of the line through (3, 1) which is parallel to the line with slope-intercept equation $y = 4x + 2$.

6.11 Find the classical slope-intercept equation [Eq. (2)] of the line through $(-1, 4)$ which is orthogonal to the line through (2, 5) and $(-1, 3)$.

6.12 Find the classical slope-intercept equation [Eq. (2)] of the line through (2, 6) which is orthogonal to the line through $(-1, 4)$ and $(-1, 8)$.

REFERENCES

1. E. T. BELL, *Men of Mathematics*, Chapter 3. New York: Simon and Schuster, 1937.

2. J. R. NEWMAN (ed.), *The World of Mathematics*, Vol. 1, page 235; Vol. 4, page 2383. New York: Simon and Schuster, 1956.

See also any text on analytic geometry or on calculus and analytic geometry.

4 | Functions

We have chosen to present the main ideas of calculus in the analysis portion of this text. Calculus deals with numerical relationships, and these relationships are given by *functions*, as we shall explain in Section 1. In this chapter we discuss those aspects of function theory which we need to present the elements of calculus.

The notion of a *map* or a *function* was introduced in Chapter 1. Recall that a **map f of a set S into a set** T is a rule which assigns to each element of S an element of T. The tendency in present-day mathematics is to reserve the term *function* for a map which has as range a subset of the set **R** of real numbers, i.e., a map $f: S \rightarrow \mathbf{R}$. In our presentation of topics in analysis, we shall be chiefly interested in functions mapping a subset of some $\mathbf{R}^n$ into **R**. We shall begin by discussing how functions appear naturally in analysis.

1. FUNCTIONS

1.1 Analysis and Functions

A mathematician may say that the area of a circular region is a *function* of the radius of the circle, meaning that the area depends on and varies with this radius. If a numerical value for the radius is given, the area of the circular region is determined. For example, if the radius is 3 units, then the area is 9π square units. Similarly, the area of a rectangular region is a function of both the length and the width of the rectangle; that is, the area depends on and varies with these quantities. If the length of a rectangle is 5 units and the width is 3 units, the rectangle encloses a region which has an area of 15 square units.

The study of how one numerical quantity Q depends on and varies with other numerical quantities is one of the major concerns of analysis. A rule which specifies the numerical value of Q for all possible values of the other quantities is often of great importance, and such a rule is of course a *function*, as we defined the term in Chapter 1.

Example 1.1 For a nonnegative real number r, the area of a circular region of radius r units is πr^2. The *circular area function* $f: \mathbf{R}^+ \rightarrow \mathbf{R}$ is the rule which says that to each nonnegative number r there is assigned the number πr^2. The equation $A = f(r) = \pi r^2$ expresses the area of the circular region as a function of its radius. ‖

147

Example 1.2 For nonnegative numbers l and w, the area of the region enclosed by a rectangle of length l and width w is lw. The *rectangular area function* $g\colon \mathbf{R}^+ \times \mathbf{R}^+ \to \mathbf{R}$ is the rule which says that to each pair (l, w) of nonnegative numbers there is assigned the number lw. The equation $A = g((l, w)) = lw$ expresses the area of the rectangular region as a function of its length and width. $\|$

In our introduction to analysis we shall be interested chiefly in functions whose domain is a subset of some Euclidean space $\mathbf{R}^n$ and whose range is contained in $\mathbf{R}$, that is, with maps of a subset of some $\mathbf{R}^n$ into $\mathbf{R}$. Such a map is a *function of n real variables*. If f has a subset D of $\mathbf{R}^n$ as domain, it is usual to let $f(x_1, \ldots, x_n)$ be the element assigned to $(x_1, \ldots, x_n) \in D$ by the function f. Thus in Example 1.2, we would use "$g(l, w)$" in place of "$g((l, w))$", dropping one pair of parentheses. For, $\mathbf{R}$, $\mathbf{R}^2$, and $\mathbf{R}^3$, it is traditional to use x, y, and z in place of x_1, x_2, and x_3, respectively.

How can you describe for someone a particular function f which you have in mind? You must specify the domain D of the function, and you must describe, in some fashion, what $f(x)$ is to be for each $x \in D$. This may be quite hard to do for some functions. Even if it is easy to describe $f(x)$ for each $x \in D$, to actually *compute* $f(c)$ for a certain $c \in D$ may be hard. For example, let n distinct points be marked on a circle and let all chords joining pairs of these points be drawn. (The points are to be such that no three chords meet at a point inside the circle.) Let $f(n)$ be the number of pieces into which the region enclosed by the circle is divided by these chords. We have defined a function f, and Example 5.3 of Chapter 1 shows that $f(1) = 1$, $f(2) = 2$, $f(3) = 4$, $f(4) = 8$, $f(5) = 16$, and $f(6) = 31$. Although f is a well-defined function with domain $\mathbf{Z}^+$, you might have some difficulty in finding $f(200)$. Frequently, a mathematical problem can be phrased in the form, "For this function f and this element c in its domain, find $f(c)$." Such a problem may be very difficult to solve.

1.2 Polynomial Functions

The functions with which we shall be chiefly concerned are the *polynomial functions*. We assume that you are familiar with the concept of a *polynomial in x with coefficients in $\mathbf{R}$* from your high school algebra. The expressions $x + 1$, x^2, and $x^3 + 4x + 3$ are such polynomials in x, and in general

$$a_n x^n + a_{n-1} x^{n-1} + \cdots + a_1 x + a_0,$$

where each $a_i \in \mathbf{R}$, is a *polynomial in x with coefficients a_i in $\mathbf{R}$*. Each such polynomial gives rise to a *polynomial function* $f\colon \mathbf{R} \to \mathbf{R}$ defined by

$$f(c) = a_n c^n + a_{n-1} c^{n-1} + \cdots + a_1 c + a_0$$

for $c \in \mathbf{R}$.

Example 1.3 For the polynomial function f given by the polynomial

$$2x^2 - x + 1,$$

we have $f(0) = 1, f(1) = 2$, and $f(-2) = 11$. ‖

We can also consider polynomials in the n symbols $x_1, x_2, \ldots, x_n$. Such a polynomial gives rise to a polynomial function f mapping $\mathbf{R}^n$ into $\mathbf{R}$. It is traditional in elementary calculus to use x, y, and z in place of x_1, x_2, and x_3, respectively. We shall often follow this convention.

Example 1.4 Consider the polynomial $x^2 - 2xy + z - 1$ in x, y, and z. If f is the corresponding polynomial function mapping $\mathbf{R}^3$ into $\mathbf{R}$, we have $f(0, 0, 0) = -1, f(1, 0, 1) = 1$, and $f(2, 1, 1) = 0$. ‖

When working with polynomial functions, questions of notation pose a real problem and are the subject of much controversy. No matter what notation an author uses, he can be sure that a sizable portion of the mathematical community will dislike it. The root of the difficulty lies in the fact that in analysis it is traditional, and often very convenient, to use the same notation for three different concepts, namely,

1) a polynomial in x,

2) the function given by the polynomial,

3) the number assigned by the polynomial function to a number x in the domain.

For example, the *polynomial* $x^2 + 2x + 3$ in x determines a *polynomial function* $f \colon \mathbf{R} \to \mathbf{R}$, traditionally also denoted by "$x^2 + 2x + 3$", which assigns to a number $x \in \mathbf{R}$ the *number* $f(x) = x^2 + 2x + 3$. With such duplication of notation, one has to judge from the context which of these three alternatives is intended. While other notations have been proposed, none has been generally accepted. We shall be traditional in this matter so that the student can read other (traditional) books easily. Although it might seem that such a multitude of meanings for one symbol would lead to terrifying confusion, in practice one has no more difficulty in choosing the correct contextual meaning than in selecting the correct meaning for the spoken words *to*, *too*, and *two* in conversation; the words sound alike. We illustrate with an example.

Example 1.5 The symbol "3" has an infinite number of meanings. For each $n \in \mathbf{Z}^+$, the *constant polynomial* 3 gives rise to the *constant function* $3 \colon \mathbf{R}^n \to \mathbf{R}$ which maps each point of $\mathbf{R}^n$ into the *number* 3. These functions are different for different values of n since they have different domains, so it is really incorrect to say, "Consider the constant function 3," unless the domain is clear from the context. ‖

1.3 Fractional Powers

A map of a set S into a set T must assign to each $x \in S$ *exactly one* element of T. It is often desirable to take a square root of a real number in $\mathbf{R}^+$; we would like to have "$\sqrt{x}$" denote a *function*. You may be accustomed to thinking of $\sqrt{4}$ as ± 2. If $\sqrt{x}$ is to be a function, we must select *just one* of these values, $+2$ or -2, to be assigned to 4 by our square root function. For this reason, it is customary in mathematics to let $\sqrt{a}$ be the *positive* square root of a for $a \in \mathbf{R}^+$. Of course, $\sqrt{0} = 0$, and the square root of a negative number is not a real number. We shall consider $\sqrt{x}$ to be the function, with domain the set D of nonnegative real numbers, which assigns to each $a \in D$ the *nonnegative* square root of a. Thus we have $\sqrt{4} = 2$, $\sqrt{9} = 3$, etc. The comments made above regarding the multiple uses of polynomial notation pertain to the notation "$\sqrt{x}$" also.

In general, if $P(x_1, \ldots, x_n)$ is any polynomial in $x_1, \ldots, x_n$, we let $\sqrt{P(x_1, \ldots, x_n)}$ be the function mapping the subset

$$D = \{(x_1, \ldots, x_n) \in \mathbf{R}^n \mid P(x_1, \ldots, x_n) \geq 0\}$$

into $\mathbf{R}$ whose value at $(a_1, \ldots, a_n) \in D$ is obtained by taking the *nonnegative* square root of $P(a_1, \ldots, a_n)$.

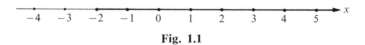

Fig. 1.1

Example 1.6 The function $\sqrt{x + 2}$ maps $D = \{x \in \mathbf{R} \mid x + 2 \geq 0\}$ into $\mathbf{R}$. If we let this function be f, we have $f(2) = 2$, $f(7) = 3$, and $f(-1) = 1$, while $f(-3)$ is undefined since $-3 \notin D$. The set D is sketched in Fig. 1.1. ‖

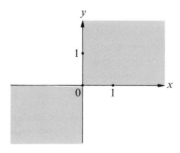

Fig. 1.2

Example 1.7 The function $\sqrt{xy}$ maps $D = \{(x, y) \in \mathbf{R}^2 \mid xy \geq 0\}$ into $\mathbf{R}$. If g is this function, we have $g(1, 1) = 1$, $g(-1, -1) = 1$, and $g(1, 4) = 2$, while $g(-1, 2)$ is undefined. The domain of g is shaded in Fig. 1.2. ‖

Recall that another notation for $\sqrt{x}$ is "$x^{1/2}$". Generalizing the above discussion in the obvious fashion, for any polynomial $P(x_1, \ldots, x_n)$ and $r/s \in \mathbf{Q}^+$, we have a function $(P(x_1, \ldots, x_n))^{r/s}$ mapping a subset D of $\mathbf{R}^n$ into $\mathbf{R}$ which assigns to $(a_1, \ldots, a_n) \in D$ the number $(\sqrt[s]{P(a_1, \ldots, a_n)})^r$. If s is odd, the domain of this function is all of $\mathbf{R}^n$, while if s is even, the domain is $D = \{(x_1, \ldots, x_n) \in \mathbf{R}^n \mid P(x_1, \ldots, x_n) \geq 0\}$. We shall not have much occasion to use these functions.

EXERCISES

1.1 Express the volume V of a cube as a function of the length x of an edge of the cube.

1.2 Express the volume V of a cylinder as a function of the radius r of the cylinder and the length l of the cylinder.

1.3 Express the area A of a circular region as a function of the perimeter s of the circle.

1.4 Express the volume V of a box with square base as a function of the length x of an edge of the base and the area A of one side of the box.

1.5 Let f be the function mapping $\mathbf{R}^2$ into $\mathbf{R}$ given by $f(x, y) = x^2 + xy$, and let g be the function mapping $\mathbf{R}$ into $\mathbf{R}$ given by $g(x) = 2x - 1$. Compute each of the following, if it is defined.

a) $f(0, 0)$ b) $f(0, 1)$ c) $f(1, 0)$ d) $g(0, 0)$
e) $f(2)$ f) $g(2)$ g) $f(2, -3)$ h) $g(2, -3)$

1.6 Let f and g be as in Exercise 1.5. Compute each of the following, if it is defined.

a) $f(g(2))$ b) $g(f(2))$ c) $g(g(g(0)))$ d) $f(f(0))$
e) $g(f(1, 2))$ f) $f(g(2, -3))$ g) $f(g(0), g(2))$

1.7 Find the domain and range of each of the following functions mapping a subset of $\mathbf{R}$ into $\mathbf{R}$.

a) $\sqrt{x + 1}$ b) $\sqrt{x^2 + 1}$ c) $\sqrt{x^2 - 1}$ d) $\sqrt{2x + 3}$

1.8 Find the domain and range of each of the following functions mapping a subset of $\mathbf{R}^2$ into $\mathbf{R}$. Sketch the domain.

a) $\sqrt{x^2 + y^2}$ b) $\sqrt{x - y}$ c) $\sqrt{1 - x^2 - y^2}$

1.9 Sketch the domain in $\mathbf{R}^2$ of each of the following functions mapping a subset of $\mathbf{R}^2$ into $\mathbf{R}$.

a) $\sqrt[3]{xy}$ b) $(xy)^{3/4}$ c) $y^{2/5}$ d) $(1 - xy)^{4/3}$

1.10 Evaluate each of the functions in Exercise 1.9 at the point $(1, -1)$, if the point is in the domain.

1.11 Describe the domain in $\mathbf{R}^4$ of each of the given functions mapping a subset of $\mathbf{R}^4$ into $\mathbf{R}$. Give a geometric description where possible.

a) $x_1 + 2x_2x_4$ b) $\sqrt{x_4}$
c) $(4 - x_1^2 - x_2^2 - x_3^3 - x_4)^{2/3}$ d) $(x_1^2 + x_2^2 + x_3^3 + x_4^4 - 16)^{3/4}$

2. THE ALGEBRA OF FUNCTIONS

In this section we shall deal only with functions which map a set S into the set $\mathbf{R}$ of real numbers. For two such functions $f\colon S \to \mathbf{R}$ and $g\colon T \to \mathbf{R}$, we can define the **sum function** $f + g$, the **difference function** $f - g$, and the **product function** fg mapping $S \cap T$ into $\mathbf{R}$ in the obvious way: namely, for $x \in S \cap T$, let

$$(f + g)(x) = f(x) + g(x) \qquad \text{and} \qquad (f - g)(x) = f(x) - g(x),$$

while

$$(fg)(x) = f(x)g(x).$$

Note that these new functions can only be defined on $S \cap T$, where both f and g are simultaneously defined. These concepts are very natural, as the next example shows.

Example 2.1 If f and g are both polynomial functions, then $f + g, f - g$, and fg are the functions obtained by taking the usual sum, difference, and product of the polynomials. For example, if $f\colon \mathbf{R} \to \mathbf{R}$ and $g\colon \mathbf{R} \to \mathbf{R}$ are given by $f(x) = x + 1$ and $g(x) = x^2 - 2$, then

$$(f + g)(x) = x^2 + x - 1,$$
$$(f - g)(x) = -x^2 + x + 3,$$
$$(fg)(x) = x^3 + x^2 - 2x - 2. \;\|$$

A *quotient function* f/g has to be handled with more care. Let us make a formal definition.

Definition. Let $f\colon S \to \mathbf{R}$ and $g\colon T \to \mathbf{R}$ be two functions and let $D = \{x \in S \cap T \mid g(x) \neq 0\}$. Then $f/g\colon D \to \mathbf{R}$ is defined by

$$\frac{f}{g}(x) = \frac{f(x)}{g(x)},$$

and is the **quotient of f by g**.

Example 2.2 If $f\colon \mathbf{R} \to \mathbf{R}$ and $g\colon \mathbf{R} \to \mathbf{R}$ are the polynomial functions given by $f(x) = x + 2$ and $g(x) = x^2 - 4$, then f/g has domain $\{x \in \mathbf{R} \mid x \neq \pm 2\}$, and

$$\frac{f}{g}(1) = \frac{f(1)}{g(1)} = \frac{1 + 2}{1 - 4} = -1. \;\|$$

In our work with analysis we shall deal primarily with *rational functions*.

Definition. A *rational function* is a quotient of two polynomial functions.

Once again, it is traditional to use the same notation for the quotient of the polynomials, the corresponding rational function, and the number assigned by the function to an element $(x_1, \ldots, x_n)$ of the domain. The meaning of the symbols is given by the context. We conclude with some more examples.

Example 2.3 The rational function

$$\frac{x^2 + y}{xy}$$

maps the subset $\{(x, y) \in \mathbf{R}^2 \mid x \neq 0, y \neq 0\}$ of $\mathbf{R}^2$ into $\mathbf{R}$. $\parallel$

Example 2.4 Without further information, it is understood from the context that the polynomial $3x + 1$ in the denominator of the quotient

$$\frac{x^2 + xy}{3x + 1}$$

corresponds to the polynomial function $g: \mathbf{R}^2 \to \mathbf{R}$ given by $g(x, y) = 3x + 1$, since the symbol "y" appears in the polynomial in the numerator. Of course, further information might show that both the numerator and denominator should be regarded as polynomial functions mapping $\mathbf{R}^3$ into $\mathbf{R}$. $\parallel$

Example 2.5 The rational function

$$\frac{3x^2 + 6xy}{3}$$

mapping $\mathbf{R}^2$ into $\mathbf{R}$ is the same as the polynomial function $x^2 + 2xy$ mapping $\mathbf{R}^2$ into $\mathbf{R}$, for it assigns the same number to each point $(x, y) \in \mathbf{R}^2$. $\parallel$

Example 2.6 The rational function

$$\frac{3x^2 + 6xy}{3x}$$

mapping $\mathbf{R}^2$ into $\mathbf{R}$ is *not* the same as the polynomial function $x + 2y$, for $x + 2y$ has $\mathbf{R}^2$ as domain, and

$$\frac{3x^2 + 6xy}{3x}$$

has $\{(x, y) \in \mathbf{R}^2 \mid x \neq 0\}$ as domain. The functions agree at points (x, y) in $\mathbf{R}^2$, where $x \neq 0$. $\parallel$

EXERCISES

2.1 To define the concepts of sum, difference, product, and quotient of functions, why did we restrict ourselves to functions mapping a set S into the set $\mathbf{R}$, rather than allowing any map of a set S into any set T?

2.2 Let f, g, and h mapping $\mathbf{R}$ into $\mathbf{R}$ be given by $f(x) = 2x + 1, g(x) = x^2 + 1$, and $h(x) = x^3 - x$. Evaluate each of the following, if it is defined.

a) $(f + g)(0)$ b) $(f - g)(2)$ c) $(gh)(-1)$ d) $\dfrac{f}{g}(0)$ e) $\dfrac{g}{h}(0)$

2.3 For the functions in Exercise 2.2, give the domain of each of the following functions.

a) $f + g$ b) $\dfrac{g}{f}$ c) $\dfrac{f}{g}$ d) $\dfrac{f}{h}$

2.4 Describe the domain and range of each of the following rational functions mapping a subset of $\mathbf{R}$ into $\mathbf{R}$.

a) $\dfrac{x^2}{1}$ b) $\dfrac{2x}{1}$ c) $\dfrac{2x}{x}$ d) $\dfrac{x^3}{x}$

e) $\dfrac{1}{x^2 + 4}$ f) $\dfrac{1}{x^2 - 4}$

2.5 Describe the domain of each of the following rational functions, from the available information.

a) $x^2 + 2xy$ b) $\dfrac{x^2 + y}{z}$ c) $\dfrac{x + 2y}{y^2 + z^2 + 1}$

d) $\dfrac{x^2}{xyz}$ e) $\dfrac{xy}{z - 1}$

2.6 In set-theoretic notation, let

$$f = \{(a, 3), (b, 5), (c, 7)\} \quad \text{and} \quad g = \{(a, 1), (b, 0), (c, 2)\}.$$

Give each of the following functions in set-theoretic notation.

a) $f + g$ b) fg c) $\dfrac{f}{g}$

2.7 Sketch the domain of each of the given functions mapping a subset of $\mathbf{R}^2$ into $\mathbf{R}$.

a) $(\sqrt{xy})(\sqrt{1 - x^2 - y^2})$ b) $(\sqrt{1 - x^2 - y^2})(\sqrt{x^2 + y^2 - 1})$ c) $\dfrac{\sqrt{xy}}{y}$

3. GRAPHS OF FUNCTIONS

There are several ways to picture or describe the behavior of a function. The technique most frequently used in classical analysis is to draw the *graph of the function*. While we shall be dealing chiefly with functions mapping a subset of $\mathbf{R}$ or $\mathbf{R}^2$ into $\mathbf{R}$, it is just as easy for us to define the graph of any map of a set S into a set T.

Definition. Let f be a map of a set S into a set T. The **graph of f** is the subset

$$\{(s, t) \in S \times T \mid f(s) = t\}$$

of the Cartesian product $S \times T$.

In terms of our discussion in Article 7.2 of Chapter 1, where we defined a function set-theoretically as a certain set of ordered pairs, the graph of a function *is* the function.

Example 3.1 Let $f: \mathbf{R} \to \mathbf{R}$ be the polynomial function $x^2 + 1$. Since $f(2) = 5$, we see that $(2, 5)$ is an element of the graph of f. Similarly, $(0, 1)$, $(3, 10)$, and $(1, 2)$ are points of the graph of f. ‖

For a function $f: \mathbf{R} \to \mathbf{R}$, its graph

$$\{(x, y) \in \mathbf{R}^2 \mid f(x) = y\}$$

is a subset of $\mathbf{R}^2$. It is often useful to sketch this subset of $\mathbf{R}^2$ in a figure, as illustrated in the following examples.

Example 3.2 Let f be the polynomial function $2x + 1$. The graph of f is $\{(x, y) \in \mathbf{R}^2 \mid 2x + 1 = y\}$. Our work in real analytic geometry shows that this set is a hyperplane of $\mathbf{R}^2$, that is, a line. This line is pictured in Fig. 3.1. ‖

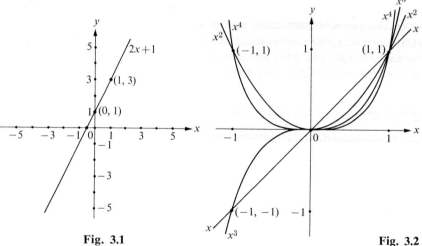

Fig. 3.1　　　　　　　　　　　　　　　　　　　　　　Fig. 3.2

Example 3.3 It is useful to know the appearance of the graph of each of the functions given by the monomials x, x^2, x^3, x^4, ... The first few of these graphs are shown in Fig. 3.2. All these graphs contain the point $(1, 1)$. Note in particular that the larger the value of n, the faster the graph of x^n climbs for $x > 1$, and the closer the graph stays to the x-axis for $0 < x < 1$. The behavior of x^n for $x < 0$ is similar to the behavior for $x > 0$, except that if n is odd, then x^n is negative for $x < 0$. ‖

Example 3.4 The graph of the rational function $1/x$ is shown in Fig. 3.3. Note that since 0 is not in the domain of the function, there is no point of the y-axis on the graph. ‖

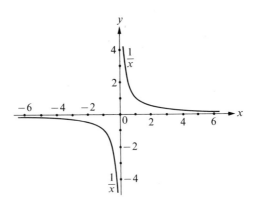

Fig. 3.3

For a function f mapping a subset of $\mathbf{R}$ into $\mathbf{R}$, the graph of f is the set of all points $(x, y) \in \mathbf{R}^2$ whose coordinates satisfy the equation $y = f(x)$. The notation "$y = f(x)$" is often used in conventional calculus texts in discussions of functions. One says that x is the *independent variable*, meaning that x may be chosen to be any element of the domain of f. On the other hand, y is the *dependent variable*, for while y might be any element of the range of f, the value of y *depends* on the value of x. Such a function f is a *function of one real variable*. A function g mapping a subset of $\mathbf{R}^2$ into $\mathbf{R}$ so that $z = g(x, y)$ is a *function of the two independent real variables x and y*, while z is the *dependent variable*, etc.

For a function $f: \mathbf{R}^2 \to \mathbf{R}$, the graph of f is the set

$$\{((x, y), z) \in \mathbf{R}^2 \times \mathbf{R} \mid f(x, y) = z\}.$$

From the point $((x, y), z) \in \mathbf{R}^2 \times \mathbf{R}$, you can obtain, in the obvious natural way, the point $(x, y, z) \in \mathbf{R}^3$. Consequently, the graph of such a function f is often sketched as a subset of $\mathbf{R}^3$.

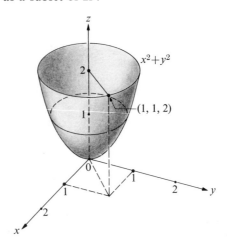

Fig. 3.4

Example 3.5 Let $f\colon \mathbf{R}^2 \to \mathbf{R}$ be the polynomial function $x^2 + y^2$ of two real variables. Then $((0, 0), 0)$, $((1, 1), 2)$, and $((1, -2), 5)$ are all points of the graph of f. Figure 3.4 indicates the subset of $\mathbf{R}^3$ corresponding in a natural way to the graph of f. The points $(0, 0, 0)$, $(1, 1, 2)$, and $(1, -2, 5)$ are in this subset of $\mathbf{R}^3$. ‖

In a full course in analytic geometry and calculus, the student is usually drilled in graphing techniques. Since it is our intention to emphasize concepts rather than techniques, we shall conclude with some simple exercises.

EXERCISES

3.1 Each of the following sets is the graph of a function. (In set-theoretic terminology, *the set is the function.*) Give the domain and the range of the function.

a) $\{(2, -3), (1, 1), (0, -3)\}$
b) $\{(1, 1), (2, 1), (3, 1)\}$
c) $\{((2, 0), 1), ((1, 1), -2), ((2, 1), 4)\}$
d) $\{(0, (0, 1)), (1, (1, -2)), (2, (1, 4))\}$

3.2 Each of the following sets fails to be the graph of a function. (In set-theoretic terminology, *the set fails to be a function.*) In each case, state why the set cannot be a graph.

a) $\{(2, -3), (4, -3), (0, 1), (2, 1)\}$
b) $\{(2, 0, 1), (1, 1, -2), (2, 1, 4)\}$
c) $\{(1, 3, -5), (-2, 1, 4), (3, -1, 2)\}$

3.3 A portion of the graph of a function f mapping $\mathbf{R}$ into $\mathbf{R}$ is shown in Fig. 3.5. Estimate each of the following from the graph.

a) $f(0)$ b) $f(1)$ c) $f(-1)$

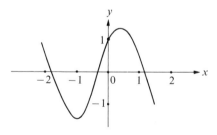

Fig. 3.5

3.4 Sketch in $\mathbf{R}^2$ the graph of each of the given polynomial functions mapping $\mathbf{R}$ into $\mathbf{R}$.

a) x b) $x + 1$ c) $x^2 - 1$ d) 3

3.5 Sketch in $\mathbf{R}^2$ the graph of each of the given rational functions mapping a subset of $\mathbf{R}$ into $\mathbf{R}$.

a) $\dfrac{1}{x - 1}$ b) $\dfrac{1}{x^2}$ c) $\dfrac{1}{x^2 - 1}$ d) $\dfrac{2x}{x}$

3.6 Sketch in $\mathbf{R}^2$ the graph of each of the following functions mapping a subset of $\mathbf{R}$ into $\mathbf{R}$.

a) $\sqrt{x}$ b) $\sqrt{x-1}$ c) $\sqrt[3]{x}$ d) $\sqrt{x^2}$

e) $\sqrt{1-x^2}$

3.7 Sketch the subset of $\mathbf{R}^3$ naturally corresponding to the graph of each of the given polynomial functions mapping $\mathbf{R}^2$ into $\mathbf{R}$.

a) $x+y$ b) x^2+y^2-1 c) $2x$

3.8 Sketch the subset of $\mathbf{R}^3$ naturally corresponding to the graph of each of the given rational functions mapping a subset of $\mathbf{R}^2$ into $\mathbf{R}$.

a) $\dfrac{1}{y^2+y^2}$ b) $\dfrac{1}{y}$ c) $\dfrac{1}{x^2}$

The remaining exercises concern another way of picturing a function. You have probably seen a map showing altitude above sea level, as illustrated in Fig. 3.6, which shows a pass between two hills. The next definition is based on this idea.

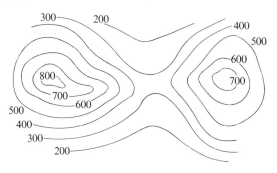

Fig. 3.6

Definition. Let $f\colon S \to T$ and let $a \in T$. The set $L_a = \{s \in S \mid f(s) = a\}$ is a *level set of f.*

Figure 3.6 shows level sets of a function which maps a portion of the plane into $\mathbf{R}$ *by assigning to each point a height above sea level.*

3.9 For each of the given polynomial functions mapping $\mathbf{R}$ into $\mathbf{R}$, find the level set L_1.

a) $3x$ b) x^2 c) $2x+3$

3.10 For each of the given polynomial functions mapping $\mathbf{R}^2$ into $\mathbf{R}$, draw a picture in $\mathbf{R}^2$ of the level set L_{-1}.

a) x^2+y^2-1 b) x^2+y^2-2 c) xy

3.11 Give a picture of the polynomial function x^2+y^2 mapping $\mathbf{R}^2$ into $\mathbf{R}$ by drawing level sets of the function, as in Fig. 3.6.

3.12 Repeat Exercise 3.11 for the polynomial function xy mapping $\mathbf{R}^2$ into $\mathbf{R}$.

4. LINEAR MAPS

4.1 Definition of a Linear Map

Among all the maps of one Euclidean space into another, the *linear maps* are the most important in analysis. Linear maps have the advantage of being easily computed. *In calculus one studies the approximation of certain maps by linear maps.*

Recall that each point $(a_1, \ldots, a_n) \in \mathbf{R}^n$ gives rise to a vector $\alpha = \prec a_1, \ldots, a_n \succ$ which we can visualize as an arrow starting at the origin and terminating at the point $(a_1, \ldots, a_n)$. For a map $f: \mathbf{R}^n \to \mathbf{R}^m$, we shall denote $f(a_1, \ldots, a_n)$ by "$f(\alpha)$" for ease in writing. The point $f(\alpha) \in \mathbf{R}^m$ again gives rise to a natural vector in $\mathbf{R}^m$; just form the pointed m-tuple of coordinates of the point $f(\alpha)$. Thus we may also view $f(\alpha)$ as a vector in $\mathbf{R}^m$. The expression "$f(\alpha) + f(\beta)$" stands for addition of vectors in $\mathbf{R}^m$, and $cf(\alpha)$ is the result of multiplying the vector $f(\alpha)$ in $\mathbf{R}^m$ by the scalar $c \in \mathbf{R}$. These conventions make it easy for us to define and discuss linear maps.

Definition. A map $f: \mathbf{R}^n \to \mathbf{R}^m$ is **linear** if for all α, $\beta \in \mathbf{R}^n$ and all $c \in \mathbf{R}$

1) $f(\alpha + \beta) = f(\alpha) + f(\beta)$,

2) $f(c\alpha) = cf(\alpha)$.

Example 4.1 Let $f: \mathbf{R}^2 \to \mathbf{R}$ be such that $f(x, y) = 2x - 3y$. Then if $\alpha = \prec a_1, a_2 \succ$ and $\beta = \prec b_1, b_2 \succ$ are any vectors in $\mathbf{R}^2$, we have

$$f(\alpha + \beta) = f(\prec a_1, a_2 \succ + \prec b_1, b_2 \succ) = f(a_1 + b_1, a_2 + b_2)$$
$$= 2(a_1 + b_1) - 3(a_2 + b_2) = (2a_1 - 3a_2) + (2b_1 - 3b_2)$$
$$= f(a_1, a_2) + f(b_1, b_2) = f(\alpha) + f(\beta).$$

Similarly, for $c \in \mathbf{R}$, we have

$$f(c\alpha) = f(c\prec a_1, a_2 \succ) = f(ca_1, ca_2) = 2ca_1 - 3ca_2$$
$$= c(2a_1 - 3a_2) = cf(a_1, a_2) = cf(\alpha).$$

Thus f is a linear map. ‖

Example 4.2 The map $f: \mathbf{R} \to \mathbf{R}$ given by $f(x) = x^2$ is not linear for $f(\prec 1 \succ + \prec 1 \succ) = f(2) = 4$, while $f(\prec 1 \succ) + f(\prec 1 \succ) = 1 + 1 = 2$. ‖

Example 4.3 Let $f: \mathbf{R} \to \mathbf{R}$ be a linear map and suppose that $f(1) = 2$. Then from the condition $f(c\alpha) = cf(\alpha)$, we have, for any $x \in \mathbf{R}$,

$$f(x) = f(x \prec 1 \succ) = xf(\prec 1 \succ) = x2 = 2x. ‖$$

Example 4.4 Let $f: \mathbf{R}^2 \to \mathbf{R}$ be a linear map and suppose that $f(1, 0) = 2$ and $f(0, 1) = -3$. The condition $f(\alpha + \beta) = f(\alpha) + f(\beta)$ shows that

$$f(1, 1) = f(\prec 1, 0 \succ + \prec 0, 1 \succ) = f(\prec 1, 0 \succ) + f(\prec 0, 1 \succ)$$
$$= 2 + (-3) = -1.$$

The condition $f(c\alpha) = cf(\alpha)$ shows that

$$f(4, 0) = f(4\langle 1, 0\rangle) = 4f(\langle 1, 0\rangle) = 4 \cdot 2 = 8.$$

Using both properties of linearity, we have, for any $(x, y) \in \mathbf{R}^2$,

$$f(x, y) = f(x\langle 1, 0\rangle + y\langle 0, 1\rangle) = xf(\langle 1, 0\rangle) + yf(\langle 0, 1\rangle)$$
$$= x \cdot 2 + y(-3) = 2x - 3y.$$

Thus f is the map of Example 4.1. ‖

The preceding examples illustrate how the values of a linear map of $\mathbf{R}^n$ into $\mathbf{R}^m$ can be computed at points of $\mathbf{R}^n$ if the values of the map at the points corresponding to the vectors

$$\beta_1 = \langle 1, 0, \ldots, 0\rangle,$$
$$\beta_2 = \langle 0, 1, \ldots, 0\rangle,$$
$$\vdots$$
$$\beta_n = \langle 0, 0, \ldots, 1\rangle$$

in $\mathbf{R}^n$ are known. We state and prove this as a theorem.

Theorem 4.1 *A linear map f of $\mathbf{R}^n$ into $\mathbf{R}^m$ is completely determined by the n values $f(\beta_1), \ldots, f(\beta_n)$ in $\mathbf{R}^m$, where β_i is as given above. If f maps $\mathbf{R}^n$ into $\mathbf{R}$ and $f(\beta_i) = a_i$, then*

$$f(x_1, \ldots, x_n) = a_1 x_1 + \cdots + a_n x_n$$

for each point $(x_1, \ldots, x_n) \in \mathbf{R}^n$. Conversely, for any $b_1, \ldots, b_n \in \mathbf{R}$, the map g of $\mathbf{R}^n$ into $\mathbf{R}$ defined by $g(x_1, \ldots, x_n) = b_1 x_1 + \cdots + b_n x_n$ is a linear map.

Proof. Let $f: \mathbf{R}^n \to \mathbf{R}^m$ be a linear map. From the two conditions $f(\alpha + \beta) = f(\alpha) + f(\beta)$ and $f(c\alpha) = cf(\alpha)$ for a linear map, it is easy to show that

$$f(c_1\alpha_1 + \cdots + c_r\alpha_r) = c_1 f(\alpha_1) + \cdots + c_r f(\alpha_r)$$

for any vectors $\alpha_i \in \mathbf{R}^n$ and scalars $c_i \in \mathbf{R}$. The vector $\langle x_1, \ldots, x_n\rangle \in \mathbf{R}^n$ can be expressed as $x_1\beta_1 + \cdots + x_n\beta_n$. Then

$$f(x_1, \ldots, x_n) = f(x_1\beta_1 + \cdots + x_n\beta_n)$$
$$= x_1 f(\beta_1) + \cdots + x_n f(\beta_n), \tag{1}$$

and we see that $f(x_1, \ldots, x_n)$ can be easily computed if $f(\beta_1), \ldots, f(\beta_n)$ are known.

Now suppose that f is a linear map of $\mathbf{R}^n$ into $\mathbf{R}$ and let $f(\beta_i) = a_i$. Then, replacing $f(\beta_i)$ by a_i, we see from Eq. (1) that

$$f(x_1, \ldots, x_n) = a_1 x_1 + \cdots + a_n x_n.$$

The proof that for $b_1, \ldots, b_n \in \mathbf{R}$ the map g of $\mathbf{R}^n$ into $\mathbf{R}$ defined by

$$g(x_1, \ldots, x_n) = b_1 x_1 + \cdots + b_n x_n$$

is indeed linear consists of a straightforward computation showing that the two conditions of the definition of a linear map are satisfied. We leave this computation to the exercises (see Exercise 4.13). ∎

4.2 The Graph of a Linear Map; Local Coordinates

If f is a linear function mapping $\mathbf{R}$ into $\mathbf{R}$ and $f(1) = a$ for $a \in \mathbf{R}$, then Theorem 4.1 shows that $f(x) = ax$ for all $x \in \mathbf{R}$. The graph of f is therefore

$$\{(x, y) \in \mathbf{R}^2 \mid ax = y\},$$

which is a hyperplane of $\mathbf{R}^2$ through the origin, that is, a line through the origin.

In general, if f is a linear map of $\mathbf{R}^n$ into $\mathbf{R}$, then by Theorem 4.1 we have $f(x_1, \ldots, x_n) = a_1 x_1 + \cdots + a_n x_n$ for some $a_i \in \mathbf{R}$. The graph of f is therefore

$$\{((x_1, \ldots, x_n), x_{n+1}) \in \mathbf{R}^n \times \mathbf{R} \mid a_1 x_1 + \cdots + a_n x_n = x_{n+1}\}.$$

If we replace $((x_1, \ldots, x_n), x_{n+1})$ by the point $(x_1, \ldots, x_n, x_{n+1})$ of $\mathbf{R}^{n+1}$, we can view the graph of f as a hyperplane in $\mathbf{R}^{n+1}$ through the origin. In particular, the graph of a linear map of $\mathbf{R}^2$ into $\mathbf{R}$ can be viewed as a plane through the origin in $\mathbf{R}^3$.

You have of course noticed the linguistic similarity of the terms *linear* and *line*. A linear map of $\mathbf{R}$ into $\mathbf{R}$ has a line of $\mathbf{R}^2$ as its graph (although, with an eye to generalization to higher dimensions, you should really view the line as a hyperplane of $\mathbf{R}^2$). It would be nice if each nonvertical line in $\mathbf{R}^2$, that is, each line in $\mathbf{R}^2$ which is the graph of a map of $\mathbf{R}$ into $\mathbf{R}$, were the graph of a linear map. Unfortunately this is not true for a line which does not pass through $(0, 0)$. However as we shall show below, we can get around this difficulty by changing our origin in $\mathbf{R}^2$ to lie on the line.

Let (h, k) be any point in $\mathbf{R}^2$ on a given nonvertical line. We introduce *local coordinates* at (h, k) by taking coordinate axes parallel to the x- and y-axes but with (h, k) as origin, as shown in Fig. 4.1. We shall call our new axes "dx- and dy-axes," as indicated in the figure. (The reason for this

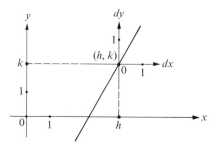

Fig. 4.1

notation will be explained in the next chapter.) In practice, we are usually only interested in working near (h, k), and this is the reason for the term *local coordinates* (in that locality) as opposed to the **global coordinates** given by the x- and y-axes.

Example 4.5 If $(h, k) = (2, -5)$ is chosen as origin for local dx- and dy-coordinates, then the point with global coordinates $(x, y) = (3, -3)$ has local coordinates $(dx, dy) = (1, 2)$, as shown in Fig. 4.2. ‖

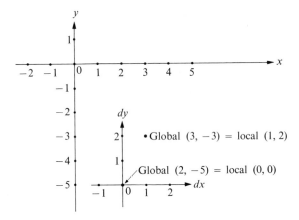

Fig. 4.2

If a line (hyperplane) in $\mathbf{R}^2$ has the equation $ax + by = c$, then we know that the vector $<a, b>$, in global coordinates, is orthogonal to the line. It is clear (see Fig. 4.3) that the vector from our local origin and with local coordinates $<a, b>$ is also orthogonal to the line. Thus the equation of the line *in local coordinates* is

$$a\, dx + b\, dy = 0 \qquad \text{or} \qquad dy = -\frac{a}{b}\, dx,$$

if $b \neq 0$, since the line passes through the local coordinate origin. By Theorem 4.1, such an equation corresponds to a linear function in terms of local dx- and dy-coordinates.

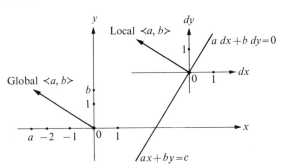

Fig. 4.3

We have shown that by choosing a suitable local origin, every non-vertical line in $\mathbf{R}^2$ can be viewed as the graph of a linear function. With an eye to generalizations to higher dimensions where a linear function mapping $\mathbf{R}^n$ into $\mathbf{R}$ has as graph a *hyperplane* in $\mathbf{R}^{n+1}$ through the origin, it would be nice if a *linear map* were called a "*hyperplanar map*." Then a "hyperplanar map" of $\mathbf{R}^n$ into $\mathbf{R}$ would always have a hyperplane in $\mathbf{R}^{n+1}$ as graph, and, conversely, most hyperplanes in $\mathbf{R}^{n+1}$ would be the graph of a "hyperplanar map," at least with respect to suitable local coordinates. Unfortunately, it is hard to get even mathematicians to agree to change established terminology in such a radical fashion.

EXERCISES

4.1 Let f be a linear map of $\mathbf{R}$ into $\mathbf{R}$ such that $f(2) = -4$. Compute the following.

a) $f(1)$ b) $f(4)$ c) $f(-2)$

4.2 Let f be a linear map of $\mathbf{R}$ into $\mathbf{R}^2$ such that $f(2) = (3, -2)$. Compute the following.

a) $f(1)$ b) $f(4)$ c) $f(-2)$

4.3 Let f be a linear map of $\mathbf{R}^2$ into $\mathbf{R}$ such that $f(1, 0) = -1$ and $f(0, 1) = 5$. Compute the following.

a) $f(-2, 1)$ b) $f(2, 4)$

4.4 Let f be a linear map of $\mathbf{R}^2$ into $\mathbf{R}^2$ such that $f(1, 0) = (2, -3)$ and $f(0, 1) = (-5, 2)$. Compute the following.

a) $f(-2, 1)$ b) $f(2, 4)$

4.5 Let f be a linear map of $\mathbf{R}^2$ into $\mathbf{R}^3$ such that $f(2, 0) = (-2, 4, 2)$ and $f(4, 2) = (8, -8, 4)$. Compute the following.

a) $f(1, 0)$ b) $f(0, 1)$ c) $f(2, 3)$

4.6 Show that the monomial function x^3 mapping $\mathbf{R}$ into $\mathbf{R}$ is not linear. [*Hint:* If $f(x) = x^3$, compute $f(\prec 1 \succ + \prec 1 \succ)$.]

4.7 Show that the map $\sqrt{x^2}$ of $\mathbf{R}$ into $\mathbf{R}$ is not linear. [*Hint:* If $f(x) = \sqrt{x^2}$, compute $f(-1 \prec 1 \succ)$.]

4.8 Show that the polynomial function $x + 3y + 2$ mapping $\mathbf{R}^2$ into $\mathbf{R}$ is not linear. [*Hint:* Argue as indicated by the hint of Exercise 4.6.]

4.9 Is the map f of $\mathbf{R}^2$ into $\mathbf{R}$ given by $f(x, y) = 0$ for all $(x, y) \in \mathbf{R}^2$ a linear map? Why?

4.10 Is the map f of $\mathbf{R}^3$ into $\mathbf{R}$ given by $f(x, y, z) = 1$ for all $(x, y, z) \in \mathbf{R}^3$ a linear map? Why?

4.11 Is the map f of $\mathbf{R}^3$ into $\mathbf{R}^4$ given by $f(x, y, z) = (0, 0, 0, 0)$ for all (x, y, z) in $\mathbf{R}^3$ a linear map? Why?

4.12 Show that any linear map of $\mathbf{R}^n$ into $\mathbf{R}^m$ maps the origin of $\mathbf{R}^n$ into the origin of $\mathbf{R}^m$.

4.13 Complete the proof of Theorem 4.1 by showing that for any $b_1, \ldots, b_n \in \mathbf{R}$, the map g of $\mathbf{R}^n$ into $\mathbf{R}$ defined by $g(x_1, \ldots, x_n) = b_1x_1 + \cdots + b_nx_n$ is a linear map.

4.14 Let the point $(-1, 3)$ in $\mathbf{R}^2$ be chosen as origin for local coordinates. Find the local dx- and dy-coordinates of each of the following points (given in global coordinates) of $\mathbf{R}^2$. Sketch in a figure like Fig. 4.2.

a) $(0, 0)$ b) $(-3, 2)$ c) $(0, 1)$ d) $(-1, -1)$

4.15 Let the point $(-1, 2, 1)$ in $\mathbf{R}^3$ be chosen as origin for local coordinates. Find the global x-, y-, and z-coordinates for the points with the given local dx-, dy-, and dz-coordinates.

a) $(0, 0, 0)$ b) $(0, 1, 2)$ c) $(-1, -3, 3)$

4.16 Find a point to serve as origin for local coordinates so that the line in $\mathbf{R}^2$ with equation $2x + 3y = 6$ becomes the graph of a linear function f mapping $\mathbf{R}$ into $\mathbf{R}$, with respect to the local coordinates. (Many answers are possible.) Give the equation of the graph in terms of the local dx- and dy-coordinates.

4.17 Repeat Exercise 4.16 for the hyperplane in $\mathbf{R}^3$ with equation $3x + 4y + z = 12$.

5. LIMITS AND CONTINUITY

5.1 The Naive Approach

The notion of a *limit* lies at the heart of classical analysis. In this section we shall explain what is meant by the *limit of a function at a point*. In the next two chapters we shall use this concept to develop the main ideas of differential and integral calculus. The notion of limit is concerned with the study of the behavior of a function near a point. Properties of a function near a point are **local properties**, as opposed to **global properties of the function** in its whole domain.

We introduce the notion of the limit of a function at a point via examples. The first example illustrates one of the uses of a limit: to describe the behavior of a function near a point not in its domain.

Example 5.1 Let f be the rational function

$$\frac{x^2 + x}{x}$$

mapping $\{x \in \mathbf{R} \mid x \neq 0\}$ into $\mathbf{R}$. The graph of f is shown in Fig. 5.1. This function assigns the same number to each nonzero element of $\mathbf{R}$ as is assigned by the polynomial function $x + 1$. The functions

$$\frac{x^2 + x}{x} \quad \text{and} \quad x + 1$$

differ only in their domains. As the graph in Fig. 5.1 shows, for numbers x close to, but different from 0, the number $(x^2 + x)/x$ is close to 1. We say

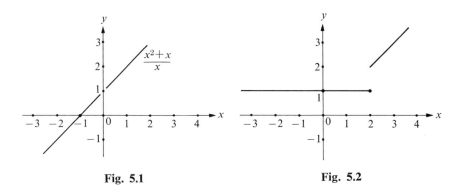

Fig. 5.1   Fig. 5.2

that *the limit of* $(x^2 + x)/x$ *as x approaches* 0 *is* 1, written

$$\text{``}\lim_{x \to 0} \frac{x^2 + x}{x} = 1\text{''}. \;\Vert$$

Example 5.2 Let $g: \mathbf{R} \to \mathbf{R}$ be defined by

$$g(x) = \begin{cases} 1 & \text{for} \quad x \le 2, \\ x & \text{for} \quad x > 2. \end{cases}$$

The graph of g is shown in Fig. 5.2. This function has an abrupt jump at 2. As x gets close to 2, but is different from 2, there is no single number which $g(x)$ approaches. If x is close to 2 but less than 2, $g(x)$ is close to 1, while if x is close to 2 but greater than 2, $g(x)$ is close to 2. In this case, *there is no limit of g as x approaches* 2. $\Vert$

    With these examples to guide us, we shall give an intuitive definition of the limit of a function at a point. A second, more precise definition, is given at the end of this section, in case your instructor wishes to discuss the concept carefully.

    We are concerned with the value of $f(x)$ for x near a point a but not equal to a. Indeed, we do not even require that a be in the domain of f; that is, $f(a)$ may not be defined. However, we do need to have points of the domain of f as close to a as we wish, so that we can discuss $f(x)$ for x as close to a as we wish. A point $a = (a_1, \ldots, a_n) \in \mathbf{R}^n$ is a *cluster point of the domain D of a function f* mapping a subset of $\mathbf{R}^n$ into $\mathbf{R}^m$ if every n-ball with center at a contains points of D other than a. For example, 1 is a cluster point of the half-open interval $]1, 2]$ of $\mathbf{R}$, although $1 \notin \;]1, 2]$. Of course, 2 is also a cluster point of $]1, 2]$. If a is a cluster point of the domain D of a function f, then you can talk about $f(x)$ for $x \in D$ as close to a as you please.

    Definition. Let f map a subset of $\mathbf{R}^n$ into $\mathbf{R}^m$ and let $a = (a_1, \ldots, a_n)$ be a cluster point of the domain of f. The *limit of f at a is* $c \in \mathbf{R}^m$ if $f(x)$

is very close to c for all $x = (x_1, \ldots, x_n)$ different from a but sufficiently close to a. The notation is

$$\text{``}\lim_{x \to a} f = c\text{''}.$$

Note that for the case in which a is in the domain of f, our definition does not require that $f(a) = c$ in order for $\lim_{x \to a} f = c$. In finding the limit at a, we are concerned with what happens for x near a but *different from a*. Let us give another example.

Example 5.3 Let $f: \mathbf{R} \to \mathbf{R}$ be defined by

$$f(x) = \begin{cases} x & \text{for} \quad x < 0, \\ 1 & \text{for} \quad x = 0, \\ x^2 & \text{for} \quad x > 0. \end{cases}$$

The graph of f is shown in Fig. 5.3. For this function f, we have

$$\lim_{x \to 0} f = 0,$$

although $f(0) = 1$. ‖

The comparison of $f(a)$ with $\lim_{x \to a} f$ concerns the *continuity* of f at a. For a function f mapping $\mathbf{R}$ into $\mathbf{R}$, to say that f is *continuous at* $a \in \mathbf{R}$ means roughly that the graph of f is not broken above the point a; that is, we can trace the graph as it goes above a with a pencil without lifting it. Thus the function f whose graph is shown in Fig. 5.3 is not continuous at 0, but is it continuous at 1.

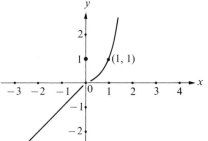

Fig. 5.3

Definition. A function f mapping a subset of $\mathbf{R}^n$ into $\mathbf{R}^m$ is ***continuous at a cluster point*** a in its domain if $\lim_{x \to a} f$ exists and is $f(a)$, that is, $\lim_{x \to a} f = f(a)$. (A function is also defined to be continuous at every point in its domain which is not a cluster point of the domain.) If f is continuous at all points in its domain, then f is ***continuous.***

It can be shown that *every rational function is continuous.* We shall assume this without proof. Most of the functions which arise naturally from a physical situation are continuous. For example, if a body is being heated, its temperature at a point is a continuous function of time.

Example 5.4 The function g of Example 5.2 whose graph is shown in Fig. 5.2 has a *discontinuity at* 2, but is continuous at all other points. ‖

Example 5.5 It makes no sense to ask whether the rational function $(x^2 + x)/x$ of Example 5.1 is continuous at 0, since 0 is not in the domain of the function. The function is continuous at every point in its domain, so it is a continuous function. ‖

*5.2 The Precise Approach

We conclude with an attempt to develop, from our intuitive definition, a more precise definition for the limit of a function at a point. In our intuitive definition of a limit, we used the phrase "$f(x)$ is very close to c." What does this mean? Different people may have different ideas about what is meant by "very close." Suppose that you want $f(x)$ to be within a distance $\epsilon > 0$ of c. (The real number ϵ may be 100, 1, $\frac{1}{2}$, 0.00000001, or any other number in $\mathbf{R}^+$, depending on how close you want $f(x)$ to be to c. The symbol "ϵ" is *always* used to denote this desired degree of closeness to c.) If $\lim_{x \to a} f = c$, then all points x sufficiently close to a, that is, within some distance $\delta > 0$ of a, but different from a, must be mapped to within the distance ϵ of c. (The symbol "δ" is *always* used to denote a number in $\mathbf{R}^+$ giving the necessary degree of closeness to a.) These numbers ϵ and δ are illustrated in Fig. 5.4 for a function mapping $\mathbf{R}$ into $\mathbf{R}$. In general, if you decide to make ϵ smaller, you will have to make δ smaller also; that is, the δ you need depends on the ϵ with which you are working. To remind you of this dependency, we shall write "δ_ϵ" in place of "δ". The precise definition is as follows.

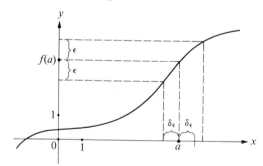

Fig. 5.4

Definition. Let f map a subset of $\mathbf{R}^n$ into $\mathbf{R}^m$ and let $a \in \mathbf{R}^n$ be a cluster point of the domain of f. The function f has *limit c at a*, written

$$\text{"}\lim_{x \to a} f = c\text{"},$$

if for each $\epsilon > 0$ in $\mathbf{R}^+$, there exists $\delta_\epsilon > 0$ in $\mathbf{R}^+$ such that for all x in the domain of f different from a, but within the distance δ_ϵ of a, the point $f(x)$ is within the distance ϵ of c.

We shall not emphasize this $(\epsilon, \delta_\epsilon)$-definition. However, we shall give one specific illustration involving ϵ and δ_ϵ.

Example 5.6 Consider the monomial function $3x$ mapping **R** into **R**. We claim that $\lim_{x\to 2} 3x = 6$. Naively, this just means that for all x sufficiently close to 2, but different from 2, $3x$ will be close to 6. Let us prove that $\lim_{x\to 2} 3x = 6$, using our $(\epsilon, \delta_\epsilon)$-definition.

Let $\epsilon > 0$ in **R** be given. We must find δ_ϵ in **R** such that if x differs from 2 by less than δ_ϵ, then $3x$ differs from 6 by less than ϵ, that is,

$$6 - \epsilon < 3x < 6 + \epsilon.$$

From this desired inequality, we obtain

$$2 - \frac{\epsilon}{3} < x < 2 + \frac{\epsilon}{3}$$

on dividing by 3. Thus we can take $\delta_\epsilon = \epsilon/3$, for if x differs from 2 by less than $\epsilon/3$, that is, if $2 - \epsilon/3 < x < 2 + \epsilon/3$, then $6 - \epsilon < 3x < 6 + \epsilon$, so $3x$ differs from 6 by less than ϵ. Of course, δ_ϵ could also be any positive number smaller than $\epsilon/3$. ‖

EXERCISES

5.1 For each condition given below, draw the graph of a function $f: \mathbf{R} \to \mathbf{R}$ which satisfies the condition.

a) Continuous at all points but 2, and $\lim_{x\to 2} f = 3$
b) Continuous at all points but 2, and $\lim_{x\to 2} f$ undefined
c) Not continuous at -1 with $\lim_{x\to -1} f = 1$, but continuous elsewhere with $\lim_{x\to 2} f = 2$

5.2 Evaluate each of the following limits, if it exists.

a) $\lim_{x\to 0} \dfrac{x^3 + x^2 + 2}{x}$ b) $\lim_{x\to 0} \dfrac{x^3 + x^2 + 2}{x^3 + 1}$ c) $\lim_{x\to 0} \dfrac{x^4 + 2x^2}{x^3 + x}$

5.3 Evaluate each of the following limits, if it exists.

a) $\lim_{x\to 2} \dfrac{x^2}{x + 3}$ b) $\lim_{x\to 1} \dfrac{(x - 1)^2}{x - 1}$ c) $\lim_{x\to 1} \dfrac{2(x - 1)}{(x - 1)^2}$

d) $\lim_{x\to -1} \dfrac{x^2 + x}{x - 1}$ e) $\lim_{x\to -1} \dfrac{x^2 + x}{x + 1}$

5.4 Evaluate each of the following limits, if it exists.

a) $\lim_{(x,y)\to(0,0)} \dfrac{x^2 + y^2 + 1}{xy}$ b) $\lim_{(x,y)\to(0,0)} \dfrac{x^2 + y^2 + 1}{xy + 2}$ c) $\lim_{(x,y)\to(0,0)} \dfrac{y}{x}$

d) $\lim_{(x,y)\to(0,0)} \dfrac{y^2}{x}$ e) $\lim_{(x,y)\to(0,0)} \dfrac{y^2 + xy + 2y}{y}$

5.5 Is the function $f\colon \mathbf{R} \to \mathbf{R}$ defined by

$$f(x) = \begin{cases} \dfrac{x^2 - 9}{x - 3} & \text{for } x \neq 3, \\ 6 & \text{for } x = 3 \end{cases}$$

continuous? Why?

5.6 Is the function $f\colon \mathbf{R} \to \mathbf{R}$ defined by

$$f(x) = \begin{cases} \dfrac{4x^2 - 2x^3}{x - 2} & \text{for } x \neq 2, \\ 8 & \text{for } x = 2 \end{cases}$$

continuous? Why?

*5.7 Using the $(\epsilon, \delta_\epsilon)$-definition, prove that the polynomial function $5x + 1$ mapping $\mathbf{R}$ into $\mathbf{R}$ is continuous at -3. [*Hint:* Follow the pattern of the argument of Example 5.6.]

*5.8 Using the $(\epsilon, \delta_\epsilon)$-definition, prove that the constant function 3 mapping $\mathbf{R}$ into $\mathbf{R}$ is continuous at 50.

*5.9 Students often have trouble understanding the $(\epsilon, \delta_\epsilon)$-definition of a limit. We are convinced that the trouble is rooted in logic. The definition uses both the *universal quantifier* (for each) and the *existential quantifier* (there exists), for the definition states that *for each $\epsilon > 0$ there exists $\delta_\epsilon > 0$*. This exercise deals with this problem in logic.

a) Negate the statement

For each $\epsilon > 0$, there exists $\delta_\epsilon > 0$.

That is, write a statement synonymous with "It is not true that for each $\epsilon > 0$, there exists $\delta_\epsilon > 0$," without just saying "It is not true that ..."

b) Negate the statement

For each apple blossom, there exists an apple.

c) Study your answer to part (a) in light of your answer to part (b), and decide whether it is correct.

d) Describe what one must do to show that $\lim_{x \to a} f \neq c$.

*5.10 Prove from the $(\epsilon, \delta_\epsilon)$-definition that for the function g of Example 5.2, $\lim_{x \to 2} g \neq \frac{1}{2}$. [*Hint:* See Exercise 5.9.]

REFERENCES

Almost any text on calculus contains the basic material we have covered on functions, except the material in Section 4. This material can be found in texts on linear algebra.

5 | Differential Calculus

Calculus was developed by Sir Isaac Newton (1642–1727), an English mathematician, and independently and at about the same time by Gottfried Wilhelm Leibniz (1646–1716), a German mathematician. We shall introduce you to calculus in this portion of the text; we know of no other topic in all mathematics which presents new ideas and techniques so quickly and gives such powerful applications so easily. We live in a changing world, and calculus is the mathematics one uses to study quantities which are changing. Indeed, Newton developed calculus to help him in his study of motion. In this chapter we shall present the main ideas of differential calculus; we shall take up integral calculus in the following chapter.

1. LOCAL LINEAR APPROXIMATION OF FUNCTIONS

1.1 The Problem

In the preceding chapter we mentioned that linear functions are important since they can easily be computed. In this section we shall show how a function can sometimes be approximated locally, that is, near a point, by a linear function. We shall restrict ourselves to functions of one real variable, i.e., to functions mapping a subset of $\mathbf{R}$ into $\mathbf{R}$.

Let f be a function mapping $\mathbf{R}$ into $\mathbf{R}$. Suppose we know the value $f(x_0)$ of f at $x_0 \in \mathbf{R}$, and wish to find the value of f at a point near x_0. For example, if f is the monomial function x^3, we know that $f(2) = 8$, and perhaps we would like to find $f(2.05)$. While $(2.05)^3$ is not too hard to compute directly, the following technique, which we shall develop in terms of a general function f, is often very useful for approximating $f(x)$ for x near a point x_0 such that $f(x_0)$ is known.

Let $y_0 = f(x_0)$, and let us set up a local coordinate system for $\mathbf{R}^2$ at the point $(h, k) = (x_0, y_0)$. Recall that we accomplish this by taking the point (x_0, y_0) as origin in the new local coordinate system, and taking a dx-axis parallel to the x-axis and a dy-axis parallel to the y-axis (see Fig. 1.1). The term *local coordinates* is used since we are only interested in studying f near x_0, that is, for small values of dx and dy. The symbols "dx" and "dy" come from a classical notation in calculus, which we shall discuss later.

Let us try to approximate f locally, near x_0, by a linear function. Remember that the graph of a linear function mapping $\mathbf{R}^n$ into $\mathbf{R}$ can be

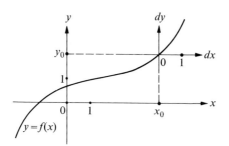

Fig. 1.1

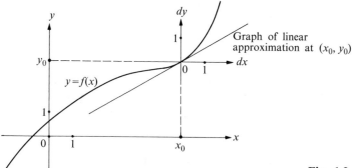

Graph of linear
approximation at (x_0, y_0)

Fig. 1.2

viewed as a hyperplane in $\mathbf{R}^{n+1}$ through the origin. In particular, a linear map of $\mathbf{R}$ into $\mathbf{R}$ has as graph a hyperplane of $\mathbf{R}^2$, that is, a line in $\mathbf{R}^2$, through the origin. Pictorially, working near the point (x_0, y_0), our new origin in local coordinates, we wish to try to approximate the graph of f by a line through this point (x_0, y_0). It is clear from Fig. 1.2 that the best linear approximation of the graph of f at (x_0, y_0) is given by *the line tangent to the graph at this point.* We mention at once that unless the function f behaves very nicely near x_0, this tangent hyperplane may not exist. A function which is not continuous at x_0, as illustrated in Fig. 1.3, has no nice local linear approximation at x_0. Also, a function f whose graph has a sharp point at (x_0, y_0) has no nice local linear approximation at x_0 (see Fig. 1.4). If the tangent line to the graph of f at (x_0, y_0) is vertical, then this line itself is not the graph of a linear function.

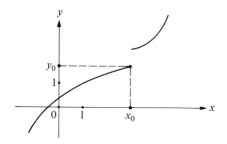

Fig. 1.3

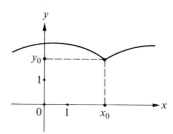

Fig. 1.4

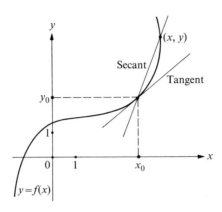

Fig. 1.5

1.2 The Derivative

Suppose that the graph of f does have a tangent hyperplane (line) at (x_0, y_0). If this tangent hyperplane is not a vertical line, then by Theorem 4.1 of the preceding chapter, the hyperplane has an equation of the form

$$dy = a\, dx$$

in terms of local coordinates. We want to find the value of a.

Since the point with local coordinates $(dx, dy) = (1, a)$ is on this tangent line, we see that the vector $\langle 1, a \rangle$ gives the direction of the line. As indicated in Fig. 1.5, the direction of the tangent line to the graph is the limiting direction of secants of the graph through (x_0, y_0) and points (x, y) on the graph, as the points (x, y) get closer and closer to (x_0, y_0). The direction of such a secant joining (x_0, y_0) and (x, y) is the direction of the vector $\langle x - x_0, y - y_0 \rangle$. Working in terms of our local coordinates, we let

$$x = x_0 + dx,$$

as shown in Fig. 1.6. Then

$$y = f(x) = f(x_0 + dx),$$

and the point (x, y) on the graph is

$$(x_0 + dx, f(x_0 + dx)).$$

The vector $\langle x - x_0, y - y_0 \rangle$ then becomes

$$\langle dx, f(x_0 + dx) - f(x_0) \rangle.$$

This vector can be viewed as an arrow emanating from our local origin and going to the point with local coordinates $(dx, f(x_0 + dx) - f(x_0))$, as shown in Fig. 1.6. Multiplying by $1/dx$, we see that the vector

$$\left\langle 1, \frac{f(x_0 + dx) - f(x_0)}{dx} \right\rangle$$

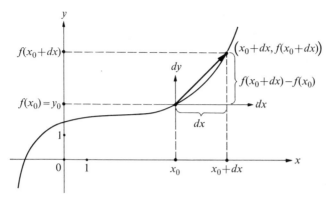

Fig. 1.6

is also directed along the secant line. The direction of our desired vector $\langle 1, a \rangle$ is the limiting direction of this vector along the secant as dx approaches 0. Thus we try to find a by attempting to evaluate

$$\lim_{dx \to 0} \frac{f(x_0 + dx) - f(x_0)}{dx}.$$

The expression

$$\text{``}\frac{f(x_0 + dx) - f(x_0)}{dx}\text{''}$$

is classically known as the **difference quotient of f at** x_0, and should be viewed as a function of the independent variable dx. This difference quotient is the change in $f(x)$ from x_0 to $x_0 + dx$, divided by the corresponding change dx in x. The limit we desire is simply that of a function of the one independent variable dx, as dx approaches 0; we discussed this situation in the preceding chapter.

> **Definition.** Let f map an open interval $]c, d[$ into **R** and let $x_0 \in]c, d[$. The limit
>
> $$\lim_{dx \to 0} \frac{f(x_0 + dx) - f(x_0)}{dx},$$
>
> if it exists, is the **derivative of f at the point** x_0, and is denoted by "$f'(x_0)$".

The direction vector $\langle 1, a \rangle$ of the tangent line to the graph of f at (x_0, y_0) is then $\langle 1, f'(x_0) \rangle$, and the equation $dy = a\, dx$ of this tangent hyperplane, in terms of local coordinates, is

$$dy = f'(x_0)\, dx. \tag{1}$$

This equation gives the local linear approximation to f at x_0.

In many books our dx of the preceding definition is denoted by "Δx", which always stands for a *change in x*. In general, in calculus texts, a notation "$\Delta thing$" stands for a change in the *thing*. Our local coordinate value dx measures the change in x from our local coordinate origin where $x = x_0$ and $dx = 0$.

In the chapter on analytic geometry we said that for a vector $\langle 1, a \rangle$ along a line in $\mathbf{R}^2$, the number a is classically called *the slope of the line*. The slope measures the amount the line climbs for each unit it traverses horizontally. As we said before, we consider slopes to be of little significance since they do not generalize to higher dimensions. In most calculus texts, however, it is strongly emphasized that for $f\colon \mathbf{R} \to \mathbf{R}, f'(x_0)$ *is the slope of the tangent line to the graph of f at* (x_0, y_0).

Example 1.1 Let f mapping $\mathbf{R}$ into $\mathbf{R}$ be the monomial function x^2, and let us find $f'(1)$. We have

$$f'(1) = \lim_{dx \to 0} \frac{f(1 + dx) - f(1)}{dx} = \lim_{dx \to 0} \frac{(1 + dx)^2 - (1)^2}{dx}$$

$$= \lim_{dx \to 0} \frac{1 + 2\,dx + (dx)^2 - 1}{dx} = \lim_{dx \to 0} \frac{2\,dx + (dx)^2}{dx}$$

$$= \lim_{dx \to 0} (2 + dx) = 2.$$

Thus the direction of the tangent line to the graph of x^2 at the point $(1, 1)$ is given by $\langle 1, 2 \rangle$. The linear approximation to x^2 at 1, in terms of local coordinates, is given by

$$dy = 2\,dx. \; \|$$

Example 1.2 Let f be the polynomial function x^3, and let us find $f'(x_0)$ for any point $x_0 \in \mathbf{R}$. We have

$$f'(x_0) = \lim_{dx \to 0} \frac{f(x_0 + dx) - f(x_0)}{dx}$$

$$= \lim_{dx \to 0} \frac{(x_0 + dx)^3 - x_0^3}{dx}$$

$$= \lim_{dx \to 0} \frac{x_0^3 + 3x_0^2\,dx + 3x_0(dx)^2 + (dx)^3 - x_0^3}{dx}$$

$$= \lim_{dx \to 0} \frac{3x_0^2\,dx + 3x_0(dx)^2 + (dx)^3}{dx}$$

$$= \lim_{dx \to 0} (3x_0^2 + 3x_0\,dx + (dx)^2) = 3x_0^2.$$

Thus $f'(x_0) = 3x_0^2$, so $f'(2) = 12, f'(4) = 48$, etc. $\|$

Recall that our purpose is to try to approximate $f(x)$ near x_0, using a local linear approximation to f. If dx is close to 0, then $x_0 + dx$ is close to

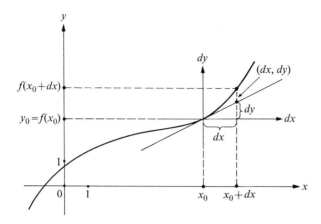

Fig. 1.7

x_0, and our desired linear approximation to $f(x_0 + dx)$ is given by the height of the tangent line in Fig. 1.7. Thus for sufficiently small values of dx, we see from Fig. 1.7 that

$$f(x_0 + dx) \approx f(x_0) + dy = f(x_0) + f'(x_0)\, dx, \qquad (2)$$

where the symbol "$\approx$" is read "is approximately equal to." *Equation (2) gives our desired approximation to f by a linear function at points $x_0 + dx$ near x_0, corresponding to values of dx near zero.* Note that, while dx is regarded as a change in x, the quantity dy is *not* the change Δy in $y = f(x)$, but is rather the approximation to Δy given by the change in the height of the *tangent line to the graph.* The notations "*dx*" and "*dy*" were introduced by Leibniz, one of the founders of calculus, who called them **differentials**.

Example 1.3 Let us estimate $(2.05)^3$ by approximating the monomial function x^3 near 2 by a linear function. If we let f be the function x^3, then Example 1.2 shows that $f'(2) = 3 \cdot 2^2 = 12$. Thus, taking $dx = 0.05$, we see from Eq. (2) that

$$(2.05)^3 = f(2.05) \approx f(2) + f'(2)\, dx = 8 + 12(0.05) = 8.6.$$

A sketch of the graph of x^3 shows that the tangent line at $(2, 8)$ lies under the graph. Therefore, this linear approximation 8.6 should be a little less than the actual value of $(2.05)^3$. Actually, $(2.05)^3 = 8.615125$. ‖

EXERCISES

1.1 For each of the following functions, form the difference quotient at the given point.

a) x^2 at 1   b) x^2 at 0   c) $\dfrac{1}{x}$ at 2

d) $\sqrt{x}$ at 4   e) $\sqrt{2x + 3}$ at 3   f) x^2 at x_0

1.2 Let f be the monomial function x^2 mapping **R** into **R**.

a) Form the difference quotient of f at 2.
b) Find $f'(2)$.
c) Find a linear estimate for $f(2.01)$.
d) Compare your answer in (c) with the actual value of $f(2.01)$.

1.3 Let f be the monomial function x^2 mapping **R** into **R**.

a) Form the difference quotient of f at -1.
b) Find $f'(-1)$.
c) Find a linear approximation to $f(-1.05)$. [*Hint:* The value of dx is *negative.*]
d) Compare your answer in (c) with the actual value of $f(-1.05)$.

1.4 Let f be the monomial function x^2. Find $f'(x_0)$ for any point $x_0 \in \mathbf{R}$.

1.5 Let f be the function $\sqrt{x}$.

a) Form the difference quotient of f at 4.
b) Find $f'(4)$. [*Hint:* Multiply the top and bottom of the difference quotient by $\sqrt{4 + dx} + \sqrt{4}$ to evaluate the limit.]
c) Find a linear approximation to $f(3.98)$. [*Hint:* Note that the value of dx is *negative.*]

1.6 Given that f is the polynomial function $x^2 + 2x$, find $f'(x_0)$ at any point $x_0 \in \mathbf{R}$.

1.7 Given that f is the rational function $1/x$, find $f'(x_0)$ at any point $x_0 \neq 0$ in **R**.

1.8 Estimate $(4.99)^3$ by taking a linear approximation to x^3 near 5. [*Hint:* Use the result of Example 1.2, which gives the derivative of x^3.]

1.9 Let f mapping **R** into **R** be the polynomial function $4x + 3$.

a) Discuss the accuracy of the linear approximation to the function at any point $x_0 \in \mathbf{R}$.
b) Without computing the limit of the difference quotient, give $f'(x_0)$ for any $x_0 \in \mathbf{R}$. [*Hint:* Argue geometrically from the graph.]

1.10 Let f be the polynomial function $x^3 + x^2$ mapping **R** into **R**.

a) Find $f'(x_0)$ for any point $x_0 \in \mathbf{R}$.
b) Compare the answer to part (a) with Example 1.2 and Exercise 1.4. Make a conjecture.

1.11 Let $f: \mathbf{R} \to \mathbf{R}$ be such that $f'(x_0)$ exists for all $x_0 \in \mathbf{R}$. Equation (1) gives the equation of the tangent hyperplane at (x_0, y_0) in terms of local dx- and dy-coordinates. Give the equation of the tangent hyperplane at (x_0, y_0) in x- and y-coordinates. [*Hint:* The vector $<1, f'(x_0)>$ is directed along the line, so $< -f'(x_0), 1>$ is orthogonal (perpendicular) to the hyperplane.]

1.12 a) Find the equation of the tangent hyperplane to the graph of the monomial function x^3 mapping **R** into **R** at the point $(4, 64)$, in terms of local coordinates at the point $(4, 64)$. [*Hint:* Use the result of Example 1.2.]
b) Repeat part (a), but find the equation of the hyperplane in terms of the x- and y-coordinates. [*Hint:* See Exercise 1.11.]

1.13 Find the xy-equation of the tangent hyperplane (line) to the graph of the rational function $1/x$ at the point $(1, 1)$. [*Hint:* See Exercises 1.7 and 1.11.]

1.14 Find parametric equations for the tangent line to the graph of the monomial function x^3 at the point $(2, 8)$.

1.15 Find parametric equations for the tangent line to the graph of the polynomial function $2x + 3$ at the point $(1, 5)$.

1.16 By examining the graphs of the monomial functions x^n for $n > 1$, find the derivative of these functions at 0.

2. THE DERIVATIVE AS A RATE OF CHANGE

2.1 The Derivative as a Function

In the preceding section we defined the concept of the derivative of a function mapping an open interval $]c, d[$ into **R** at a point $x_0 \in]c, d[$. Let us extend this definition somewhat.

> **Definition.** Let f be a function mapping a subset D of **R** into **R** such that every point of the domain D of f is inside some open interval contained wholly within D. The function f is **differentiable at** $x_0 \in D$ if
>
> $$f'(x_0) = \lim_{dx \to 0} \frac{f(x_0 + dx) - f(x_0)}{dx}$$
>
> exists. The function is **differentiable** if it is differentiable at every point in D. If f is differentiable, then the function $f': D \to$ **R** whose value at $x_0 \in D$ is $f'(x_0)$ is the **derivative of** f.

Example 2.1 From Example 1.2, we see that the derivative of the monomial function x^3 is the function $3x^2$. ‖

2.2 Rate of Change

The derivative has a useful interpretation as a rate of change. Indeed, this interpretation is responsible for the wide application of differential calculus. Let us illustrate.

Consider a moving car whose speed may vary. The distance the car has gone is a function of the time it has been traveling. If "s" denotes distance and "t" denotes time, then $s = g(t)$ for some function g. We may measure the distance s in miles and the time t in hours. Suppose, for example, that the car is traveling so that $s = g(t) = t^3$ (this is a very unusual car). Since we know how far the car has gone at every instant, it seems reasonable that we should be able to find the speed of the car at any instant. Speed is the *rate of change of distance per unit change of time.* Let us try to find the speed of our car after two hours, that is, when $t = 2$.

When $t = 2$, we have $s = 8$, and when $t = 4$, we have $s = 64$. Thus during the two-hour time interval between $t = 2$ and $t = 4$, the car has traveled $64 - 8 = 56$ mi. To have traveled 56 mi in two hours, the car

dt	$g(2 + dt)$	$g(2 + dt) - g(2)$	$\dfrac{g(2 + dt) - g(2)}{dt}$
2	64	56	28
1	27	19	19
0.5	15.625	7.625	15.250
0.1	9.261	1.261	12.610
0.01	8.120601	0.120601	12.0601
0.001	8.012006001	0.012006001	12.006001

Fig. 2.1

must have had an average speed of $\frac{56}{2} = 28$ miles per hour (mph). Similarly, at time $t = 3$, the car has traveled 27 mi. During the one-hour interval between $t = 2$ and $t = 3$, the car has thus gone $27 - 8 = 19$ mi, and has averaged 19 mph.

At time $t = 2 + dt$, the car has traveled

$$g(2 + dt) = (2 + dt)^3 \text{ mi.}$$

In the time interval of dt hr between $t = 2$ and $t = 2 + dt$, the car has gone

$$g(2 + dt) - g(2) \text{ mi.}$$

To have gone $g(2 + dt) - g(2)$ mi in dt hr, its *average* speed must have been

$$\frac{g(2 + dt) - g(2)}{dt} \text{ mph.}$$

Values of this average speed over shorter and shorter time intervals, that is, for smaller and smaller values of dt, are given in the table in Fig. 2.1.

We would expect that the average speed of 12.006001 mph given in the last entry for the time interval between $t = 2.000$ and $t = 2.001$ hr would be very close to the actual speed of the car at time $t = 2$. *The instantaneous speed at time $t = 2$, the reading of the speedometer needle at that instant, is the limiting value of these average speeds as dt approaches zero.* Thus

$$(\textit{instantaneous speed at } t = 2) = \lim_{dt \to 0} \frac{g(2 + dt) - g(2)}{dt}.$$

The expression

$$\lim_{dt \to 0} \frac{g(2 + dt) - g(2)}{dt}$$

should look familiar. By similar reasoning, we see that for $f: \mathbf{R} \to \mathbf{R}$, *the difference quotient*

$$\frac{f(x_0 + dx) - f(x_0)}{dx}$$

is the average rate of change of $f(x)$ *per unit change of* x *between* x_0 *and* $x_0 + dx$, and

$$f'(x_0) = \lim_{dx \to 0} \frac{f(x_0 + dx) - f(x_0)}{dx}$$

is the instantaneous rate that $f(x)$ *is changing per unit change in* x *at* x_0. The great power of calculus lies in its usefulness in the study of changing systems, as opposed to static systems.

Example 2.2 We saw in the preceding section that the derivative of the monomial function x^3 is $3x^2$. Whether the symbol "x" or "t" is used is immaterial. Thus the derivative of the distance function g given by $g(t) = t^3$ in our illustration above is g', where $g'(t) = 3t^2$. When $t = 2$, we see that the instantaneous speed, $g'(2)$, is $3 \cdot 2^2 = 12$ mph, as we would have guessed from the table in Fig. 2.1. ∥

2.3 The Leibniz Notation

We now introduce a classical notation of calculus. This notation is due to Leibniz. Leibniz's notation is very useful in applications and in helping us remember certain formulas of calculus, as we shall indicate in the following section. In the preceding section we saw that if $y = f(x)$, then a nonvertical tangent hyperplane (line) to the graph of f at the point (x_0, y_0) has the equation

$$dy = f'(x_0)\, dx$$

in local coordinates. For a point with local coordinates (dx, dy) on the hyperplane, we therefore have

$$\frac{dy}{dx} = f'(x_0)$$

if $dx \neq 0$. The notation

$$``\frac{dy}{dx}"$$

is Leibniz's classical notation for the derivative f'. If $y = f(x)$, the notation "y'" is also often used for the derivative f'.

Rather than writing

$$``\frac{dy}{dx}(x_0)"$$

to denote $f'(x_0)$, one usually writes

$$\text{``}\frac{dy}{dx}\bigg]_{x_0}\text{''},$$

read

$$\text{``}\frac{dy}{dx}\text{ evaluated at }x_0.\text{''}$$

The notation "dy/dx" is useful in rate of change applications, where it suggests the *instantaneous rate of change of y with respect to x*. You should remember that

$$\frac{d(\text{first quantity})}{d(\text{second quantity})} = \begin{cases}\textit{instantaneous rate of change of the first}\\ \textit{quantity per unit change of the second.}\end{cases}$$

It does not make sense to ask simply for the rate of change of a quantity. A quantity can only have a rate of change *with respect to some other quantity.*
    We conclude with some examples.

Example 2.3 The volume V of a cube is given as a function of the length x of an edge by the equation $V = x^3$. By Example 1.2,

$$\frac{dV}{dx} = 3x^2$$

gives the instantaneous rate of change of the volume per unit change in the length of an edge for each value of x. ‖

Example 2.4 The area A enclosed by a circle is given as a function of the radius r by $A = \pi r^2$. Thus dA/dr gives the instantaneous rate of change of the area per unit change in the radius r for each value of r. We shall see in the next section that

$$\frac{dA}{dr} = 2\pi r. \text{ ‖}$$

Example 2.5 If the distance s traveled by a car is given as a function of time t by $s = g(t)$, then

$$v = \frac{ds}{dt} = g'$$

gives the instantaneous speed v of the car at each time t. The rate of change of speed with respect to time is *acceleration*, so dv/dt gives the acceleration of the car. ‖

Example 2.6 The profit that a manufacturer makes is a function of the amount of his product that he produces. If he produces nothing, he makes no profit. If he produces less than the demand, he can charge a good price and make a good profit. If he produces much more than the demand, he may not be able to sell all he produces at any price, and he may make very little profit, or even lose money. Thus his profit P is a function of the amount x he produces. The derivative dP/dx gives the rate of change of profit P per unit change in the amount x he produces. This **marginal profit** dP/dx is of interest in the study of economics. Economists usually use the adjective *marginal* to denote a derivative, i.e., an instantaneous rate of change. ‖

EXERCISES

2.1 Suppose that Bill and Mike are running in a race. Let $s_B = f(t)$ be the distance Bill has run and let $s_M = g(t)$ be the distance Mike has run at time t. Let distance be measured in feet and time in seconds. Give the physical interpretation of each of the following.

a) $\left.\dfrac{ds_B}{dt}\right]_{10} = 15$ b) $\left.\dfrac{ds_M}{dt}\right]_{60} = 12$ c) $\left.\dfrac{ds_B}{dt}\right]_{30} = \left(\dfrac{3}{2}\right)\left.\dfrac{ds_M}{dt}\right]_{30}$

2.2 Suppose an unusual car travels so that after t hr it has gone $3t^2 + 2t$ mi.

a) Find the average speed of the car during the two-hour time interval between $t = 3$ and $t = 5$.
b) Find the average speed of the car during the one-hour time interval between $t = 3$ and $t = 4$.
c) Find the average speed of the car during the half hour between $t = 3$ and $t = \frac{7}{2}$.
d) On the basis of the preceding parts, guess the actual speed of the car at time $t = 3$.

2.3 With reference to Exercise 2.2, find, from the definition of the derivative as a limit, the derivative of the polynomial function $3t^2 + 2t$ when $t = 3$. Compare your answer with your guess in (d) of Exercise 2.2.

2.4 With reference to Exercises 2.2 and 2.3, find the acceleration of the car as a function of time.

2.5 An equilateral triangle x units long on each side has an area A of

$$\frac{\sqrt{3}}{4} x^2 \text{ square units.}$$

Express in words the meaning of dA/dx as a rate of change.

2.6 Let $f: \mathbf{R} \to \mathbf{R}$ be the constant function 3. View f' as giving a rate of change, and find f' without using the definition of the derivative as a limit.

2.7 If a body travels a distance $s = g(t)$ in time t, the **graph of its motion** is the graph of g, that is, $\{(t, s) \in \mathbf{R}^2 \mid s = g(t)\}$.

a) If the body is stationary, what is the graph of its motion?
b) If the body is moving at a constant speed v_0, so that $ds/dt = v_0$ for all t, what is the graph of its motion?

2.8 If the length of an edge of a cube increases at a rate of 1 in./sec, find the rate of increase of volume when

a) the edge is 2 in. long,
b) the edge is 5 in. long.

[*Hint:* See Example 2.3.]

2.9 Repeat Exercise 2.8, assuming that the edge of the cube is increasing at a rate of 4 in./sec.

2.10 Suppose that when a pebble is dropped into a large tank of fluid a wave travels outward in a circular ring whose radius increases at a constant rate of 8 in./sec.

a) Find the area of the circular region enclosed by the wave two seconds after the time the pebble hits the fluid.
b) Find the rate that the area of the circular region enclosed by the wave is increasing two seconds after the time the pebble hits the fluid. [*Hint:* The necessary derivative is given in Example 2.4.]

3. COMPUTATION OF THE DERIVATIVE

If we intend just to present concepts rather than to develop techniques, we should drop our study of differential calculus for functions of one real variable at this point, and go directly to functions of several variables. Except for a few technical theorems which we shall not attempt to present, we have covered the main ideas of differential calculus for functions mapping **R** into **R**. Further work in this area in the usual calculus course consists of enlarging the stockpile of functions to include the trigonometric, exponential, and logarithmic functions, developing techniques, and learning some applications. We cannot resist the temptation to develop a little technique and to give some applications, for we know of no other branch of mathematics where such impressive applications can be given so quickly. Remember, the many applications of differential calculus arise from the fact that *the derivative allows us to study dynamic systems, that is, systems which are changing.*

3.1 The Derivative of a Polynomial

We shall devote this section to teaching you how to compute the derivative of some of the functions in the meager stockpile which we built in the last chapter. We see no particular virtue in carefully deriving these differentiation formulas for you; we shall attempt to make them seem reasonable. The derivations of the formulas can be found in any calculus text designed for a full course.

We saw in Example 1.2 of Section 1 that the derivative of the monomial function x^3 is $3x^2$. Using exactly the same technique, the binomial theorem, one can easily show that the derivative of the monomial function x^n is nx^{n-1} for all $n \in \mathbf{Z}^+$.

Example 3.1 The derivative of x^4 is $4x^3$; the derivative of x^{15} is $15x^{14}$; and the derivative of

$$x = x^1 \quad \text{is} \quad 1x^0 = 1 \cdot 1 = 1. \, \|$$

The formula

$$\frac{d(x^r)}{dx} = rx^{r-1}$$

can be shown to hold for all $r \in \mathbf{R}$, although we have not defined the function x^r except for $r \in \mathbf{Q}$. Recall from high school algebra that $x^{-n} = 1/x^n$.

Example 3.2 The derivative of

$$\frac{1}{x^2} = x^{-2} \quad \text{is} \quad -2 \cdot x^{-3} = \frac{-2}{x^3}.$$

The derivative of

$$\frac{1}{x} = x^{-1} \quad \text{is} \quad -1 \cdot x^{-2} = -\frac{1}{x^2}.$$

The derivative of

$$\sqrt{x} = x^{1/2} \quad \text{is} \quad \tfrac{1}{2}x^{-1/2} = \frac{1}{2\sqrt{x}}. \, \|$$

Recall that dy/dx gives the instantaneous rate of change of y with respect to x. If y is increasing at a rate of a units per unit change in x, then surely for any constant c, cy increases at a rate of $c \cdot a$ units per unit change in x. For example, if we double y, we double the increase, etc. This gives us the formula

$$(cf)' = c \cdot f', \quad \text{or} \quad \frac{d(cy)}{dx} = c\frac{dy}{dx}$$

for any $c \in \mathbf{R}$.

Example 3.3 The derivative of the monomial function $4x^5$ is $4(5x^4) = 20x^4. \, \|$

If y is increasing at a rate of a units per unit change in x, and z is increasing at a rate of b units per unit change in x, then surely, taking y and z together, $y + z$ is increasing at the combined rate of $a + b$ units per unit change in x. That is,

$$\frac{d(y + z)}{dx} = \frac{dy}{dx} + \frac{dz}{dx}.$$

Example 3.4 The derivative of the polynomial function $4x^3 + 5x^7$ is the sum $4(3x^2) + 5(7x^6) = 12x^2 + 35x^6$ of the derivatives of the two monomials. The derivative of

$$\frac{1}{x^2} + 2x^3 = x^{-2} + 2x^3 \quad \text{is} \quad -2(x^{-3}) + 2(3x^2) = \frac{-2}{x^3} + 6x^2. \, \|$$

Since the rate of change of a constant function is zero, that is,

$$\frac{d(c)}{dx} = 0$$

for all constant functions $c \colon \mathbf{R} \to \mathbf{R}$, we have the following theorem.

Theorem 3.1 *The derivative of the polynomial function*

$$a_n x^n + a_{n-1} x^{n-1} + \cdots + a_1 x + a_0$$

is

$$n a_n x^{n-1} + (n-1) a_{n-1} x^{n-2} + \cdots + a_1.$$

Example 3.5 For the function f where

$$f(x) = 4x^5 - 2x^2 + \frac{1}{x^3},$$

let us estimate $f(1.01)$ by taking a local linear approximation to f at 1. We easily find that

$$f(1) = 4 - 2 + 1 = 3.$$

Using local coordinates dx and dy at the point $(1, 3)$, we know from Section 1 that

$$f(1 + dx) \approx f(1) + f'(1)\, dx$$

for sufficiently small dx. We easily compute that

$$f'(x) = 20x^4 - 4x - \frac{3}{x^4},$$

so $f'(1) = 20 - 4 - 3 = 13$. Thus, taking $dx = 0.01$, we obtain

$$f(1.01) \approx 3 + 13(0.01) = 3.13. \;\|$$

Example 3.6 Let us estimate $\sqrt{98}$. If $f(x) = \sqrt{x}$, we have $f(100) = \sqrt{100} = 10$. We take local coordinates at the point $(100, 10)$, and take a linear approximation to $\sqrt{x}$, given by

$$f(100 + dx) \approx f(100) + f'(100)\, dx.$$

Now

$$f'(x) = \tfrac{1}{2} x^{-\frac{1}{2}}, \qquad \text{so} \qquad f'(100) = \tfrac{1}{20}.$$

Taking $dx = -2$, we have

$$f(98) \approx 10 + \tfrac{1}{20}(-2) = 10 - \tfrac{1}{10} = 9.9.$$

You may think that -2 is much too large a value for dx to give a good estimate. However, the graph of $\sqrt{x}$ curves so slowly at $(100, 10)$ that the tangent line is close to it for quite a way. The actual value of $\sqrt{98}$, to six decimal places, is 9.899495, so our error is only about 0.0005. $\|$

3.2 The Chain Rule

Frequently, a quantity y may depend on a quantity u which in turn depends on a quantity x. That is, $y = f(u)$ and $u = g(x)$ for some functions f and g. Then, given a value of x, one can find the corresponding value of u, and finally a value for y, so that y appears as a function h of x; that is,

$$y = h(x) = f(g(x)).$$

We would like to find dy/dx in terms of dy/du and du/dx.

Theorem 3.2 (Chain rule for differentiation). *In the situation just described, if all derivatives exist, we have*

$$\frac{dy}{dx} = \frac{dy}{du}\frac{du}{dx}.$$

Explanation. If Sam can run twice as fast as Mary, and Mary can run three times as fast as baby Jane, how many times as fast as baby Jane can Sam run? Obviously, $2 \cdot 3 = 6$ times as fast. Thus

$$\frac{d(\text{Sam})}{d(\text{Jane})} = \frac{d(\text{Sam})}{d(\text{Mary})}\frac{d(\text{Mary})}{d(\text{Jane})}.$$

This is the way to think of the chain rule, which is really obvious if the derivative is viewed as a rate of change.

The chain rule illustrates the advantage of the Leibniz notation for the derivative in helping us remember formulas. You can remember

$$\frac{dy}{dx} = \frac{dy}{du}\frac{du}{dx}$$

by remembering that the "du's cancel."

Example 3.7 Let us find the derivative of the function f where $f(x) = (4x^3 + 7x^2)^{10}$. We could raise the polynomial $4x^3 + 7x^2$ to the tenth power, and obtain one polynomial expression for f, but this is hard work. Let $y = (4x^3 + 7x^2)^{10}$ and let $u = 4x^3 + 7x^2$. Then $y = u^{10}$ and by the chain rule,

$$\frac{dy}{dx} = \frac{dy}{du}\frac{du}{dx} = 10u^9(12x^2 + 14x) = 10(4x^3 + 7x^2)^9(12x^2 + 14x).$$

Our problem is solved. ‖

Example 3.7 illustrates the formula

$$\frac{d(u^n)}{dx} = nu^{n-1}\frac{du}{dx}.$$

This formula follows at once from the chain rule, for we have

$$\frac{d(u^n)}{dx} = \frac{d(u^n)}{du}\frac{du}{dx} = nu^{n-1}\frac{du}{dx}.$$

Example 3.8 In solving Exercise 2.9, you used common sense amounting to the chain rule. Let us solve Exercise 2.9 here by using the chain rule. The length of each edge of a cube is increasing at a rate of 4 in./sec, and we want to find the rate of increase in the volume V when the edges are 2 in. long. If an edge is of length x, we have $V = x^3$. We are asked to find the rate of change of V *with respect to time*, that is, dV/dt. Now

$$\frac{dV}{dt} = \frac{dV}{dx}\frac{dx}{dt} = 3x^2\frac{dx}{dt}$$

by the chain rule. We are given that $dx/dt = 4$ in./sec. Thus when $x = 2$, we have

$$\frac{dV}{dt} = 12 \cdot 4 = 48 \text{ cu in./sec. } \parallel$$

The exercises are designed to give you limited practice in computing derivatives and in solving problems like those illustrated in this section. We are more interested in having you realize what *can* be done using calculus than in having you develop great facility in doing it.

EXERCISES

3.1 Find the derivative of each of the indicated functions.

a) x^7 b) $5x^6$ c) $3x^4 + 7x^2$

d) $\dfrac{3}{x}$ e) $\dfrac{5}{x^2} + \dfrac{1}{x}$ f) $2x^3 + \dfrac{3}{x^3}$

g) $8x^2 + 3x + 4 - \dfrac{7}{x}$ h) $3x - 7 + \dfrac{14}{x} - \dfrac{7}{x^3}$

3.2 Find the derivative of each of the indicated functions.

a) $\sqrt{x}$ b) $\dfrac{4}{\sqrt{x}}$ c) $x^2 - 2(\sqrt[3]{x})$

d) $3x^4 + 2x^{1/2} - x^{-1/3} + 7x^{-4}$

3.3 Use the formula

$$\frac{d(u^n)}{dx} = nu^{n-1}\frac{du}{dx}$$

to find the derivative of each of the indicated functions.

a) $(3x + 1)^{10}$ b) $(x^2 - 5x^3)^5$ c) $\sqrt{5x}$

d) $\sqrt{3x^2 + 1}$ e) $\dfrac{1}{\sqrt[3]{4x^3 + 9x}}$ f) $((x^2 + 4x)^3 - 9x^2)^5$

3.4 Estimate $\sqrt[3]{26}$ by taking a linear approximation to the function $x^{1/3}$.

3.5 Estimate $(4.01)^3 + 3(\sqrt{4.01})$ by taking a linear approximation to the function $x^3 + 3x^{1/2}$.

3.6 Estimate $((1.05)^3 + 1)^4$ by using a linear approximation to the function $(x^3 + 1)^4$.

3.7 Estimate $\sqrt{63} + (0.95)^5$, using two linear approximations.

3.8 Given that the radius of a circle is increasing at a constant rate of 3 in./sec, find the rate of increase of the area of the region enclosed by the circle when the radius is 4 in.

3.9 The volume enclosed by a spherical balloon is decreasing at a uniform rate of 12 cu ft/min. Find the rate of decrease of the radius when the radius is 8 ft. [*Hint:* The volume V of a sphere is given in terms of the radius r by the formula $V = \frac{4}{3}\pi r^3$.]

3.10 Ship A passes a buoy at 9:00 a.m. and continues on a northward course at a rate of 12 mph. Ship B, traveling at 18 mph, passes the same buoy on its eastward course at 10:00 a.m. the same day. Find the rate at which the distance between the ships is increasing at 11:00 a.m. that day. [*Hint:* See Fig. 3.1. You want to find dz/dt. Use the Pythagorean theorem, and differentiate with respect to t, using the chain rule.]

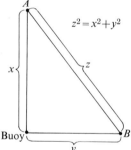

Fig. 3.1

3.11 It is not difficult to derive formulas for the derivative of the product and the quotient of differentiable functions f and g mapping subsets of $\mathbf{R}$ into $\mathbf{R}$. With

$$u = f(x) \quad \text{and} \quad v = g(x),$$

we have

$$\frac{d(uv)}{dx} = u\frac{dv}{dx} + v\frac{du}{dx} \quad \text{and} \quad \frac{d\left(\dfrac{u}{v}\right)}{dx} = \frac{v\dfrac{du}{dx} - u\dfrac{dv}{dx}}{v^2}.$$

Use these formulas to find the derivative of each of the following functions.

a) $\dfrac{x}{x+1}$

b) $\dfrac{x^2}{x^2-2}$

c) $(x-2)^3(2x+3)^2$

d) $\dfrac{x^2-3x+1}{x^2-2}$

e) $\dfrac{(x+2)^3}{(x-1)^2}$

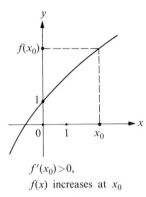

$f'(x_0)>0,$

$f(x)$ increases at x_0

Fig. 4.1

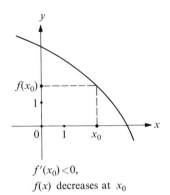

$f'(x_0)<0,$

$f(x)$ decreases at x_0

Fig. 4.2

4. EXTREMUM PROBLEMS

Frequently one wishes to maximize or minimize some quantity. For example, a manufacturer may want to maximize his profit. A builder may want to minimize his materials. A very important application of differential calculus concerns problems of this sort.

4.1 The Sign of the Derivative

Let f mapping a subset D of **R** into **R** be a differentiable function. By considering just the sign (positive, negative, or zero) of $f'(x)$ for x near $x_0 \in D$, it is possible to tell whether $f(x)$ is increasing, decreasing, changes from increasing to decreasing, or changes from decreasing to increasing at x_0 as x increases. If $f'(x_0) > 0$, then the rate of change of $f(x)$ per unit change of x at x_0 is positive, so $f(x)$ increases as x increases at x_0. Similarly, if $f'(x_0) <$ 0, then $f(x)$ decreases at x_0. These situations are shown in Figs. 4.1 and 4.2.

Suppose that $f'(x_0) = 0$. Four types of situations which frequently occur are illustrated in Figs. 4.3 through 4.6. In the case illustrated in Fig. 4.5,

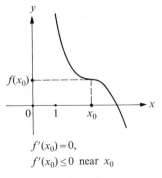

$f'(x_0)=0,$

$f'(x_0) \leq 0$ near x_0

Fig. 4.3

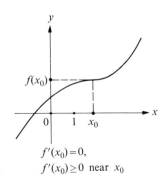

$f'(x_0)=0,$

$f'(x_0) \geq 0$ near x_0

Fig. 4.4

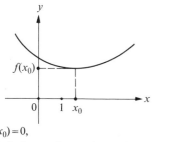

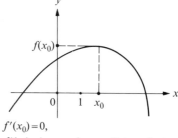

$f'(x_0) = 0,$
$f'(x_0)$ changes from <0 to >0 at x_0 $f'(x_0) = 0,$
$f'(x_0)$ changes from >0 to <0 at x_0

Fig. 4.5   **Fig. 4.6**

where $f'(x_0) = 0$ and $f'(x)$ changes from negative to positive as x increases at x_0, the function f has a *local minimum* at x_0. In Fig. 4.6, f has a *local maximum* at x_0. We summarize these ideas in a definition and a theorem. We shall not give a careful proof of the theorem; our figures should make it seem reasonable.

Definition. Let f map a subset of $\mathbf{R}$ into $\mathbf{R}$ and let x_0 be a point in the domain of f. The function f has a **local minimum at** x_0 if $f(x_0) < f(x)$ for all $x \neq x_0$ in the domain of f and in a sufficiently small open interval containing x_0. If $f(x_0) > f(x)$ for such $x \neq x_0$, then f has a **local maximum at** x_0.

Theorem 4.1 *Let the domain of f mapping a subset of $\mathbf{R}$ into $\mathbf{R}$ contain an open interval $]a, b[$ and let f be differentiable at each point of this interval. Let $x_0 \in]a, b[$.*

1) If f has a local maximum or a local minimum at x_0, then $f'(x_0) = 0$.

2) If $f'(x_0) = 0$ and $f'(x)$ changes sign from negative to positive at x_0 as x increases, then f has a local minimum at x_0.

3) If $f'(x_0) = 0$ and $f'(x)$ changes sign from positive to negative at x_0 as x increases, then f has a local maximum at x_0.

4.2 Computation of Extremum Values

To find local minima and maxima of a function f mapping a subset of $\mathbf{R}$ into $\mathbf{R}$, we may proceed as follows. Find all points $x \in \mathbf{R}$ where $f'(x) = 0$. These are candidates for points at which f has a local minimum or maximum. See whether the derivative of f changes sign at these points, and determine whether each gives a local minimum, a local maximum, or neither. Then worry about points which are not inside an open interval in which f is differentiable (if any such points exist).

We conclude with two specific applications of this theory. Further applications are given in the exercises. Let us outline a procedure for you to follow in these applications.

STEP 1. Decide what you want to minimize or maximize.

STEP 2. Express this quantity which you wish to minimize or maximize as a function f of *one* other quantity. (You may have to draw a figure and use some algebra to do this.)

STEP 3. Find all points x where $f'(x) = 0$.

STEP 4. Decide whether your desired minimum or maximum occurs at one of the points you found in Step 3. Frequently it will be clear that a minimum or maximum exists from the nature of the problem, and if Step 3 gives only one candidate, that is your answer. If there is more than one candidate, or if there are points in the domain of f which are not inside an open interval in which f is differentiable, you may have to make a further examination.

Example 4.1 Let us find two numbers whose sum is 6 and whose product is as large as possible.

STEP 1. We want to maximize the product P of two numbers.

STEP 2. If the two numbers are x and y, then $P = xy$. Since $x + y = 6$, we have $y = 6 - x$, so $P = x(6 - x) = 6x - x^2$.

STEP 3. We have

$$\frac{dP}{dx} = 6 - 2x.$$

Thus $dP/dx = 0$ when $6 - 2x = 0$, or when $x = 3$.

STEP 4. Since $dP/dx = 6 - 2x$ is greater than 0 for x less than 3 and is less than 0 for x greater than 3, we see that P has a maximum at 3.

Thus the largest value of P occurs when $x = 3$ and $y = 3$, so $xy = 9$. ‖

Example 4.2 A manufacturer of dog food wishes to package his product in cylindrical metal cans, each of which is to contain a certain volume V_0 of dog food. What should be the ratio of the height of the can to its radius in order to minimize the amount of metal, assuming that the top, bottom, and cylinder are all made from metal of the same thickness?

STEP 1. The manufacturer wishes to minimize the surface area S of the can.

STEP 2. The surface of the can consists of the circular disks at the top and bottom, and the cylinder. If the radius of the can is r and the height is h, the top and bottom disks each have area πr^2, and the cylinder has area $2\pi rh$ (see Fig. 4.7). Thus

$$S = 2\pi r^2 + 2\pi rh.$$

We would like to find h in terms of r to express S as a function of the single quantity r. The volume $V_0 = \pi r^2 h$, so

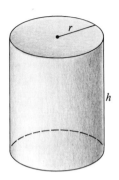

$$h = \frac{V_0}{\pi r^2}.$$

Thus

$$S = 2\pi r^2 + 2\pi r \frac{V_0}{\pi r^2} = 2\left(\pi r^2 + \frac{V_0}{r}\right).$$

STEP 3. We easily find that

$$\frac{dS}{dr} = 2\left(2\pi r - \frac{V_0}{r^2}\right).$$

Fig. 4.7

Thus $dS/dr = 0$ when $2\pi r - V_0/r^2 = 0$, so $2\pi r^3 = V_0$, and

$$r^3 = \frac{V_0}{2\pi}.$$

STEP 4. It is obvious from the nature of the problem that a minimum for S does exist. A can $\frac{1}{32}$ in. high and 6 ft across uses a lot of metal, as does a can $\frac{1}{32}$ in. in radius and a 100 ft high. Somewhere between these extremes is a can of reasonable dimensions which uses less metal. We found only one candidate in Step 3, so we don't have to look further. We were interested in the ratio h/r. Now $V_0 = \pi r^2 h$, so

$$h = \frac{V_0}{\pi r^2} \quad \text{and} \quad \frac{h}{r} = \frac{V_0}{\pi r^3}.$$

From Step 3, the least metal is used when $r^3 = V_0/2\pi$, so

$$\frac{h}{r} = \frac{V_0}{\pi \dfrac{V_0}{2\pi}} = 2. \;\|$$

    Example 4.2 illustrates the practical importance of such problems. For a cylindrical can of minimal surface area and of a given volume, we should have $h = 2r$, so the height should equal the diameter. Thus few cans on the shelves in our supermarket represent efficient packaging, assuming that the top, bottom, and cylinder are of equally expensive material. The tuna fish cans are usually too short, and the soft drink and beer cans are too high.

EXERCISES

For additional extremum problems, see any calculus text.

4.1 Find $b \in \mathbf{R}$ such that the polynomial function $x^2 + bx - 7$ has a local minimum at 4.

4.2 For $a \in \mathbf{R}$, consider the polynomial function $ax^2 + 4x + 13$.

a) Find the value of a such that the function has either a local maximum or a local minimum at 1. Which is it, a maximum or a minimum?

b) Find the value of a such that the function has either a local maximum or a local minimum at -1. Which is it, a maximum or a minimum?

4.3 For $a, b \in \mathbf{R}$, consider the polynomial function f given by $ax^2 + bx + 24$.

a) Find the ratio b/a if f has a local minimum at 2.

b) Find a and b if f has a local minimum at 2 and $f(2) = 12$.

c) Can you determine a and b so that f has a local maximum at 2 and $f(2) = 12$?

4.4 Find the maximum area a rectangle can have if the perimeter is 20 ft.

4.5 Generalize Exercise 4.4 to show that the rectangle of maximum area having a fixed perimeter is a square.

4.6 Suppose you wish to construct a cardboard box with volume 108 cu in. and with a square base and open top. Find the area of cardboard necessary. (Neglect waste in construction.)

4.7 Find two positive numbers x and y such that $x + y = 6$ and xy^2 is as large as possible.

4.8 A farmer has 1000 rods of fencing with which to fence in three sides of a rectangular pasture; a straight river will form the fourth side of the pasture. Find the dimensions of the pasture of largest area which he can fence.

4.9 Find the dimensions of the rectangle of maximum area which can be inscribed in a semicircle of radius r (see Fig. 4.8). [*Hint:* You may find it easier to maximize the square of the area. Of course, the rectangle having maximum area is the one having the square of its area maximum.]

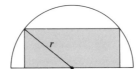

Fig. 4.8

4.10 A silo is to have the form of a cylinder capped with a hemisphere. Given that the material for the hemisphere is twice as expensive per square foot as the material for the cylinder, find the ratio of the height of the cylinder to its radius for the most economical structure. Neglect waste in construction. (The volume of a sphere of radius r is $\frac{4}{3}\pi r^3$, and the surface area is $4\pi r^2$.)

5. MOTION ON A STRAIGHT LINE

5.1 Velocity and Acceleration

Consider a body traveling in a straight path. By choosing an origin and a suitable scale, we can consider the line on which the body is moving to be a number line (see Fig. 5.1). At each time t, measured in convenient units from some instant corresponding to $t = 0$, the body is at some location s on this number line. Thus s appears as a function of the time t. The derivative ds/dt,

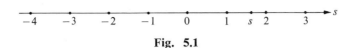

Fig. 5.1

if it exists, gives the rate of change of the location of the body, measured from the origin, per unit change in time. This rate of change is the *velocity v of the body*. Thus

$$v = \frac{ds}{dt}.$$

The rate of change of the velocity with respect to time is the *acceleration a of the body*. Therefore if the derivative of the velocity with respect to time exists, the acceleration is

$$\frac{dv}{dt} = \frac{d\left(\frac{ds}{dt}\right)}{dt}.$$

This derivative is usually denoted by

$$\text{`` } \frac{d^2s}{dt^2} \text{ ''}.$$

In general, if $y = f(x)$, then $d^n y / dx^n$, the *nth derivative of f with respect to x*, is the function obtained by differentiating f, then f', etc., until n differentiations have been performed. Thus

$$a = \frac{d^2s}{dt^2} = \text{second derivative of } s \text{ with respect to } t$$

and

$$\frac{da}{dt} = \frac{d^3s}{dt^3} = \text{rate of change of acceleration with respect to time,}$$

etc.

Example 5.1 Suppose the position s of a body traveling on a number line is given in feet at time t seconds by $s = 2t^3 - t + 2$. We can easily find the velocity v and the acceleration a of the body at any time t. We have

$$v = \frac{ds}{dt} = 6t^2 - 1 \quad \text{and} \quad a = \frac{d^2s}{dt^2} = 12t.$$

Thus after five seconds, the body has a velocity of 149 ft/sec and an acceleration of 60 ft/sec². [The units which are used to measure acceleration are *(unit of distance)/(unit of time)²*.] At time $t = 0$, our body had an initial velocity of $v_0 = -1$. The negative value of v_0 means that s was decreasing with respect to time when $t = 0$; that is, our body was moving in the negative direction on our number line at $t = 0$. ‖

5.2 Antiderivatives and Differential Equations

Often one knows the acceleration or the velocity of a body as a function of time, and wishes to find the location s of the body at time t. Suppose, for example, that you know the velocity ds/dt as a function of t. How can you then find the position s in terms of t? You must find some function which has the known ds/dt as derivative. Let us turn to the mathematics of such a situation.

> **Definition.** Let f be a function mapping a subset of $\mathbf{R}$ into $\mathbf{R}$. A function F with the same domain as f is an ***antiderivative of f*** if F' exists and $F' = f$.

Example 5.2 An antiderivative of the polynomial function $2x + 1$ is $x^2 + x$, for

$$\frac{d(x^2 + x)}{dx} = 2x + 1.$$

Note that another antiderivative of $2x + 1$ is $x^2 + x + 3$, for

$$\frac{d(x^2 + x + 3)}{dx} = 2x + 1$$

also. ‖

It is easy to see that for $n \neq -1$, an antiderivative of the monomial function x^n is

$$\frac{1}{n + 1} x^{n+1}.$$

Example 5.2 illustrates a general situation which we state in a theorem. We shall give a naive explanation of this theorem. A rigorous proof depends on a more careful study of $\mathbf{R}$.

> ***Theorem 5.1*** *Let $f: \mathbf{R} \to \mathbf{R}$ be a differentiable function. If $f' = 0$, that is, if $f'(x) = 0$ for all $x \in \mathbf{R}$, then f must be a constant function; that is, there exists $c \in \mathbf{R}$ such that $f(x) = c$ for all $x \in \mathbf{R}$. If $f: \mathbf{R} \to \mathbf{R}$ and $g: \mathbf{R} \to \mathbf{R}$ are two differentiable functions and if $f'(x) = g'(x)$ for all $x \in \mathbf{R}$, then for some $c \in \mathbf{R}$, $f(x) = g(x) + c$ for all $x \in \mathbf{R}$; that is, the functions f and g differ by a constant function.*

Explanation. If $f'(x) = 0$ for all $x \in \mathbf{R}$, then the rate of change of $f(x)$ per unit change of x is zero at each point $x_0 \in \mathbf{R}$. Thus $f(x)$ never changes at all as x changes, so if $f(x_0) = c$, then $f(x) = f(x_0) = c$ for all $x \in \mathbf{R}$.

Suppose now that $f'(x) = g'(x)$ for all $x \in \mathbf{R}$, and let $h = f - g$, so that

$$h(x) = f(x) - g(x).$$

Then

$$h'(x) = f'(x) - g'(x) = 0$$

for all $x \in \mathbf{R}$, so by the first part of the theorem, there exists $c \in \mathbf{R}$ such that $h(x) = c$ for all $x \in \mathbf{R}$. Therefore $f(x) - g(x) = c$, so $f(x) = g(x) + c$ for all $x \in \mathbf{R}$.

Example 5.3 Referring to Example 5.2, we see that any antiderivative of $2x + 1$ has the form $x^2 + x + c$, where $c \in \mathbf{R}$. The quantity c is called *an arbitrary constant*, and $x^2 + x + c$ is the *general antiderivative of* $2x + 1$. ‖

You can usually determine which antiderivative F of a function f you want if you know $F(x_0)$ for some single x_0 in the domain of F.

Example 5.4 Let f be the polynomial function $2x + 1$. Suppose you want to find an antiderivative F of f such that $F(1) = -3$. We know we must have

$$F(x) = x^2 + x + c$$

for some $c \in \mathbf{R}$. Using the "boundary condition" $F(1) = -3$, we have

$$-3 = F(1) = 1^2 + 1 + c = 2 + c.$$

Thus we must have $c = -5$, so the desired F is the polynomial function $x^2 + x - 5$. ‖

An equation which involves one or more derivatives is a **differential equation**. The basic problem of differential equations is finding *solutions* of the equations, that is, finding functions whose derivatives satisfy a differential equation. For example, in Example 5.3, we solved the differential equation

$$\frac{dy}{dx} = 2x + 1$$

and obtained the *general solution* $y = x^2 + x + c$. In Example 5.4, we found the *particular solution* $y = x^2 + x - 5$ such that $y = -3$ when $x = 1$.

5.3 A Freely Falling Body

We conclude with the standard application of these ideas to the study of a freely falling body.

Example 5.5 It is known that the acceleration of a body falling in a vacuum (i.e., with air resistance neglected) near the surface of the earth is the gravitational acceleration g. The value of g is about 32 ft/sec². Suppose that a body is initially at a height of 160 ft above the surface of the earth, and is thrown straight up with a velocity of 48 ft/sec. Let us find the velocity and height of this body at all future times t until it strikes the earth, neglecting air resistance (i.e., for a body in a vacuum).

We take as our number line a vertical line with the origin at the surface of the earth and the positive direction upward, measured in feet. From our data, we obtain the differential equation

$$\frac{d^2s}{dt^2} = -32,$$

which is valid for all t until the body hits the ground, and we also know that

$$\frac{ds}{dt}\bigg]_0 = 48 \text{ ft/sec}$$

and

$$s = 160 \text{ ft} \qquad \text{when} \qquad t = 0.$$

The negative value for the acceleration is due to the fact that the acceleration is downward, in the *negative* direction on our number line. The initial velocity is upward, in the *positive* direction, so this quantity 48 is positive. Now ds/dt is an antiderivative of d^2s/dt^2. Thus by Theorem 5.1,

$$v = \frac{ds}{dt} = -32t + c_1$$

for some $c_1 \in \mathbf{R}$. Since

$$\frac{ds}{dt}\bigg]_0 = 48,$$

we find that

$$48 = -32(0) + c_1,$$

so $c_1 = 48$. Hence

$$v = \frac{ds}{dt} = -32t + 48.$$

Since s is an antiderivative of ds/dt, we obtain

$$s = -32(\tfrac{1}{2}t^2) + 48t + c_2 = -16t^2 + 48t + c_2$$

for some $c_2 \in \mathbf{R}$. When $t = 0$, we have $s = 160$, so

$$160 = -16(0^2) + 48(0) + c_2,$$

and $c_2 = 160$. Therefore

$$s = -16t^2 + 48t + 160.$$

We have found ds/dt and s as functions of t. Of course, the equations only hold until the body reaches the surface of the earth, that is, until the positive time t when $s = 0$. Solving the equation $0 = -16t^2 + 48t + 160$, we obtain

$$t^2 - 3t - 10 = 0,$$

so $(t - 5)(t + 2) = 0$. This equation has $t = 5$ as positive solution. Thus our equations for ds/dt and s are valid for $0 \leq t \leq 5$. ‖

EXERCISES

5.1 Find the second, third, and fourth derivatives of each of the indicated functions.

a) $3x$ b) $x^5 - 3x^2$ c) $\dfrac{1}{x}$ d) $x^3 - \dfrac{3}{x^2}$

5.2 Find the most general antiderivative (with an arbtriary constant) of each of the indicated functions.

a) 3 b) $4x^2$ c) $x^3 - 2x^2$ d) $x^4 - \dfrac{7}{x^2}$

e) $3x^5 + 2x^3 - 3 + \dfrac{5}{x^3}$

5.3 Find the antiderivative F of the function $5x^4 - 4/x^2$ such that $F(2) = 4$.

5.4 Find the general solution of the differential equation

$$\frac{d^2y}{dx^2} = 12x + 6.$$

5.5 Find the particular solution of the differential equation

$$\frac{d^2x}{dt^2} = 24t^2 - 6$$

such that

$$\frac{dx}{dt}\bigg]_0 = 0 \qquad \text{and} \qquad x = 4 \qquad \text{when} \quad t = 0.$$

5.6 Find the general solution of the differential equation

$$\frac{d^3y}{dx^3} = x + \frac{1}{x^4}.$$

5.7 For the motion discussed in Example 5.5, find the maximum height of the body above the surface of the earth.

5.8 If a body is dropped from a height of 256 ft above the surface of the earth, find the time required for it to fall to the surface, neglecting air resistance.

5.9 Derive the general formulas giving the velocity v and height s at time t of a body thrown upward in a vacuum with an initial velocity v_0 from an initial height s_0 above the surface of the earth at time $t = 0$.

5.10 The acceleration of revolution for a certain spinning wheel at time t minutes is given by the formula

$$a = 12t + 4 \text{ rpm}^2.$$

Given that the speed of revolution of the wheel after ten minutes is 700 rpm, find what the speed of revolution of the wheel was after five minutes.

5.11 Give an example to show that if the domain of f is not all of $\mathbf{R}$, and if $f'(x) = 0$ for all x in the domain of f, it need not be the case that $f(x) = c$ for some $c \in \mathbf{R}$ at all points in the domain of f.

5.12 Let $y = f(x)$ give a function mapping $\mathbf{R}$ into $\mathbf{R}$ such that both dy/dx and d^2y/dx^2 exist for all $x \in \mathbf{R}$. Suppose $f'(3) = 0$.

a) Given

$$\frac{d^2y}{dx^2}\bigg]_3 > 0,$$

argue that f has a local minimum at 3. [*Hint:* Think of this second derivative as the rate of change of the first derivative, and consider Figs. 4.3 through 4.6.]

b) Given

$$\frac{d^2y}{dx^2}\bigg]_3 < 0,$$

argue that f has a local maximum at 3. [*Hint:* Proceed as in (a).]

6. PARTIAL DERIVATIVES

In this section we start to study local linear approximations of a function of more than one variable. We shall generalize our work in Section 1 to a function f mapping a subset of $\mathbf{R}^2$ into $\mathbf{R}$.

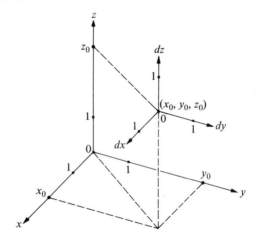

Fig. 6.1

6.1 Tangent Hyperplanes and Partial Derivatives

Let f map a subset D of $\mathbf{R}^2$ into $\mathbf{R}$, and let $(x_0, y_0) \in D$. We wish to approximate f locally, near (x_0, y_0), by a linear map. We shall assume that (x_0, y_0) is the center of some 2-ball (disk) lying wholly within D.

Let $z_0 = f(x_0, y_0)$. Recall that we can view the graph of f as a subset of $\mathbf{R}^3$ by renaming the point $((x_0, y_0), z_0) \in \mathbf{R}^2 \times \mathbf{R}$ of the graph by (x_0, y_0, z_0) in $\mathbf{R}^3$. Proceeding as in Section 1, we choose local coordinates dx, dy, and dz in $\mathbf{R}^3$ at (x_0, y_0, z_0), as indicated in Fig. 6.1. By Theorem 4.1 of the preceding chapter, the graph of a function mapping $\mathbf{R}^2$ into $\mathbf{R}$, which is linear with respect to our local coordinates, can be viewed as a hyperplane (a plane) of

$\mathbf{R}^3$ through our local origin (x_0, y_0, z_0) with an equation

$$dz = a\,dx + b\,dy.$$

We must determine the values of a and b to obtain a linear approximation to f. If we take $dx = 1$ and $dy = 0$, we see that the point with local coordinates $(1, 0, a)$ lies in the hyperplane. Similarly, the point with local coordinates $(0, 1, b)$ lies in the hyperplane. Thus the vectors $\prec 1, 0, a \succ$ and $\prec 0, 1, b \succ$, viewed as emanating from our local origin (x_0, y_0, z_0), both lie in this hyperplane.

In general, for a continuous map f of $\mathbf{R}^2$ into $\mathbf{R}$, the graph of f is some sort of a surface in $\mathbf{R}^3$. We may argue as in Section 1 that a good local linear approximation to f at (x_0, y_0) should have as graph the *tangent hyperplane* (*plane*) to this surface at (x_0, y_0, z_0). We mention at once that unless the function f has a nice smooth surface for its graph, this tangent plane need not exist. In particular, a tangent plane does not exist if f is not continuous at (x_0, y_0, z_0).

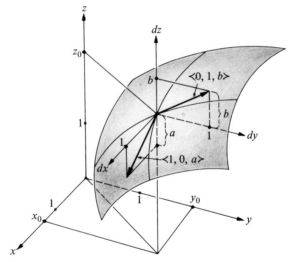

Fig. 6.2

Figure 6.2 shows a typical situation. At the point (x_0, y_0, z_0), consider the curve consisting of all points in the graph of f whose y-coordinate is y_0. This curve is the intersection of the graph and the hyperplane with equation $y = y_0$. The vector $\prec 1, 0, a \succ$ emanating from (x_0, y_0, z_0) lies in the tangent hyperplane and must be tangent to this curve. Arguing precisely as in Section 1, the direction of this vector $\prec 1, 0, a \succ$ is the limiting direction of the secants of the curve through (x_0, y_0, z_0) and $(x_0 + dx, y_0, z)$ as dx approaches zero. Thus, as in Section 1,

$$a = \lim_{dx \to 0} \frac{f(x_0 + dx, y_0) - f(x_0, y_0)}{dx}$$

if this limit exists. In a similar fashion, we find that

$$b = \lim_{dy \to 0} \frac{f(x_0, y_0 + dy) - f(x_0, y_0)}{dy}$$

if this limit exists.

Definition. Let f map a subset of $\mathbf{R}^2$ into $\mathbf{R}$ and let (x_0, y_0) be the center of some 2-ball which contains only points in the domain of f. Then

$$\lim_{dx \to 0} \frac{f(x_0 + dx, y_0) - f(x_0, y_0)}{dx},$$

if it exists, is the **partial derivative of f with respect to x at** (x_0, y_0), and is denoted by

$$``\frac{\partial f}{\partial x}\bigg]_{(x_0, y_0)} ".$$

The partial derivative

$$\frac{\partial f}{\partial y}\bigg]_{(x_0, y_0)}$$

is similarly defined. If these partial derivatives exist at every point in the domain of f, then

$$\frac{\partial f}{\partial x},$$

the **partial derivative of f with respect to x**, is the function mapping the domain of f into $\mathbf{R}$ whose value at (x_0, y_0) is

$$\frac{\partial f}{\partial x}\bigg]_{(x_0, y_0)}.$$

The partial derivative $\partial f/\partial y$ of f with respect to y is similarly defined.

Thus if the tangent hyperplane to the graph of $f: \mathbf{R}^2 \to \mathbf{R}$ exists at (x_0, y_0, z_0), its equation, in terms of local coordinates at (x_0, y_0, z_0), is

$$dz = \frac{\partial f}{\partial x}\bigg]_{(x_0, y_0)} dx + \frac{\partial f}{\partial y}\bigg]_{(x_0, y_0)} dy.$$

The equation of this tangent hyperplane in global coordinates is

$$z - z_0 = \frac{\partial f}{\partial x}\bigg]_{(x_0, y_0)} (x - x_0) + \frac{\partial f}{\partial y}\bigg]_{(x_0, y_0)} (y - y_0),$$

for this gives a hyperplane with the same (orthogonal) direction vector and passing through the point (x_0, y_0, z_0).

6.2 Computation of Partial Derivatives

We can regard

$$\frac{\partial f}{\partial x}\bigg]_{(x_0,\, y_0)}$$

as the derivative at x_0 of the function $g\colon \mathbf{R} \to \mathbf{R}$ given by $g(x) = f(x, y_0)$, that is, the function of one real variable x obtained from f by keeping $y = y_0$ and allowing only x to vary. This means that we can compute partial derivatives by using the techniques given in Section 3, as illustrated in the following examples.

Example 6.1 Let $f\colon \mathbf{R}^2 \to \mathbf{R}$ be the polynomial function $x^2 + 3xy + 2y^3$. Let us find

$$\frac{\partial f}{\partial x}\bigg]_{(1,2)} \quad \text{and} \quad \frac{\partial f}{\partial y}\bigg]_{(1,2)}.$$

Now the first of these can be viewed as the derivative at 1 of the polynomial function $g\colon \mathbf{R} \to \mathbf{R}$ obtained by setting $y = 2$ in $x^2 + 3xy + 2y^3$. That is,

$$g(x) = x^2 + 3x(2) + 2 \cdot 2^3 = x^2 + 6x + 16.$$

The derivative of $x^2 + 6x + 16$ is $2x + 6$, and the value of $2x + 6$ at $x = 1$ is 8. Hence

$$\frac{\partial f}{\partial x}\bigg]_{(1,2)} = 8.$$

Similarly, putting $x = 1$, we find that $\partial f/\partial y$ evaluated at $(1, 2)$ is the derivative at 2 of $1 + 3y + 2y^3$. Since

$$\frac{d(1 + 3y + 2y^3)}{dy} = 3 + 6y^2,$$

we obtain

$$\frac{\partial f}{\partial y}\bigg]_{(1,2)} = 3 + 6 \cdot 2^2 = 27. \;\|$$

Example 6.2 We could simplify the computation in Example 6.1 by noting that we can compute $\partial f/\partial x$ at any (x, y) by differentiating $x^2 + 3xy + 2y^3$ with respect to x only, treating y as a constant. Thus

$$\frac{\partial f}{\partial x} = \frac{\partial(x^2 + 3xy + 2y^3)}{\partial x} = 2x + 3y,$$

for

$$\frac{\partial(2y^3)}{\partial x} = 0,$$

since we think of y as a constant. Similarly,

$$\frac{\partial f}{\partial y} = \frac{\partial(x^2 + 3xy + 2y^3)}{\partial y} = 3x + 6y^2.$$

Therefore

$$\left.\frac{\partial f}{\partial x}\right]_{(1,2)} = 2 \cdot 1 + 3 \cdot 2 = 8$$

and

$$\left.\frac{\partial f}{\partial y}\right]_{(1,2)} = 3 \cdot 1 + 6 \cdot 2^2 = 27. \;\|$$

Example 6.3 For $f(x, y) = x^2 + 3xy + 2y^3$ as in Example 6.1, we find that the tangent hyperplane to the graph of f at the point $(1, 2, 23)$ is given in local coordinates by

$$dz = 8\,dx + 27\,dy \qquad \text{or} \qquad 8\,dx + 27\,dy - dz = 0.$$

Thus the vector $\langle 8, 27, -1\rangle$ is orthogonal to the tangent hyperplane. In x-, y-, and z-coordinates, therefore, the hyperplane has an equation

$$8(x - 1) + 27(y - 2) - 1(z - 23) = 0 \qquad \text{or} \qquad 8x + 27y - z = 39. \;\|$$

For a function $f\colon \mathbf{R}^3 \to \mathbf{R}$, the partial derivatives $\partial f/\partial x$, $\partial f/\partial y$, and $\partial f/\partial z$ are similarly defined. To compute $\partial f/\partial x$, we regard all variables but x as constants, and we take the derivative with respect to x. Since $\partial f/\partial x$ is in general again a function of x, y, and z, we can attempt to compute second derivatives; for example,

$$\frac{\partial\left(\frac{\partial f}{\partial x}\right)}{\partial x} = \frac{\partial^2 f}{\partial x^2}, \qquad \frac{\partial\left(\frac{\partial f}{\partial x}\right)}{\partial y} = \frac{\partial^2 f}{\partial y\,\partial x}, \qquad \text{and} \qquad \frac{\partial\left(\frac{\partial f}{\partial x}\right)}{\partial z} = \frac{\partial^2 f}{\partial z\,\partial x}.$$

We could also compute

$$\frac{\partial\left(\frac{\partial f}{\partial y}\right)}{\partial x} = \frac{\partial^2 f}{\partial x\,\partial y}.$$

It can be shown that for "nice functions," in particular, for rational functions, the order in which successive partial differentiations are performed is immaterial, so that, for example,

$$\frac{\partial^2 f}{\partial x\,\partial y} = \frac{\partial^2 f}{\partial y\,\partial x}.$$

We shall stop here and give you practice in partial differentiation in the exercises. Applications concerning the approximation of functions near a point and rate of change are taken up in Section 7.

EXERCISES

6.1 Given $f(x, y) = xy + 3y^2$, find

$$\left.\frac{\partial f}{\partial x}\right]_{(-2,3)} \quad \text{and} \quad \left.\frac{\partial f}{\partial y}\right]_{(-2,3)}.$$

6.2 Find the equation of the tangent hyperplane to the graph of the polynomial function $xy + 3y^2$ at the point $(-2, 3, 21)$ in terms of

a) local coordinates at $(-2, 3, 21)$,   b) global x-, y-, and z-coordinates.

6.3 For each of the given functions f, find $\partial f/\partial x$ and $\partial f/\partial y$.

a) xy b) $x^2 + 3y^2$ c) $xy^3 + x^2y^2$

d) $\dfrac{x}{y}$ e) $\dfrac{x^2 + 3x + 1}{y}$ f) $xy^2 + \dfrac{3x^2}{y^3}$

6.4 Verify that

$$\frac{\partial^2 f}{\partial x\, \partial y} = \frac{\partial^2 f}{\partial y\, \partial x}$$

for each of the following functions.

a) x^2y b) $x^3y^2 + \dfrac{x^2}{y}$

6.5 Verify that

$$\frac{\partial^3 f}{\partial y\, \partial x^2} = \frac{\partial^3 f}{\partial x\, \partial y\, \partial x}$$

for the function $x^3y^2 + x^2/y^3$. By definition,

$$\frac{\partial^3 f}{\partial y\, \partial x^2} = \frac{\partial\left(\dfrac{\partial^2 f}{\partial x^2}\right)}{\partial y} \quad \text{and} \quad \frac{\partial^3 f}{\partial x\, \partial y\, \partial x} = \frac{\partial\left(\dfrac{\partial^2 f}{\partial y\, \partial x}\right)}{\partial x}.$$

6.6 Let $f: \mathbf{R}^3 \to \mathbf{R}$ be the polynomial function $x^2yz + yz^3 - 3x^2y^5$. Compute each of the following.

a) $\left.\dfrac{\partial f}{\partial x}\right]_{(1,-1,0)}$ b) $\dfrac{\partial^2 f}{\partial x^2}$ c) $\dfrac{\partial^3 f}{\partial x\, \partial y\, \partial z}$

d) $\left.\dfrac{\partial^4 f}{\partial z\, \partial y\, \partial x^2}\right]_{(0,0,0)}$ e) $\dfrac{\partial^5 f}{\partial x^2\, \partial y^3}$

6.7 Find the equation of the tangent hyperplane at $(1, 1, -1, 1)$ to the 3-sphere in $\mathbf{R}^4$ with center at the origin and radius 2.

6.8 Find the equation of the tangent hyperplane at $(1, -2, -1, 1, -3)$ to the 4-sphere in $\mathbf{R}^5$ with center at the origin and radius 4.

6.9 A line is *normal to the graph of a function* if it is orthogonal to the tangent hyperplane to the graph. Find parametric equations of the line in $\mathbf{R}^3$ which is normal to the graph of the function $x^2 + 3xy + 2y^3$ at the point $(1, 2, 23)$.

6.10 With reference to Exercise 6.9, find parametric equations of the line in $\mathbf{R}^4$ which is normal to the graph of the function $x^2y - xy^2z^3$ at the point $(1, 2, -1, 6)$.

7. APPROXIMATION AND RATE OF CHANGE

7.1 Local Linear Approximation

We saw in the preceding section that a local linear approximation at (x_0, y_0) to a function $f: \mathbf{R}^2 \to \mathbf{R}$ is given in local coordinates by

$$dz = \frac{\partial f}{\partial x}\bigg]_{(x_0, y_0)} dx + \frac{\partial f}{\partial y}\bigg]_{(x_0, y_0)} dy,$$

provided that the graph of f has a tangent hyperplane at (x_0, y_0, z_0). Approximating f linearly near (x_0, y_0), just as we did in Section 1 for a function of one real variable, we have

$$f(x_0 + dx, y_0 + dy) \approx f(x_0, y_0) + dz$$

$$= f(x_0, y_0) + \frac{\partial f}{\partial x}\bigg]_{(x_0, y_0)} dx + \frac{\partial f}{\partial y}\bigg]_{(x_0, y_0)} dy$$

for dx and dy sufficiently small. Similar formulas hold for local linear approximations of a map of $\mathbf{R}^n$ into $\mathbf{R}$. For example, if g maps $\mathbf{R}^3$ into $\mathbf{R}$ with $w_0 = g(x_0, y_0, z_0)$, and if the graph of g has a tangent hyperplane at (x_0, y_0, z_0, w_0), then

$$g(x_0 + dx, y_0 + dy, z_0 + dz) \approx g(x_0, y_0, z_0) + \frac{\partial g}{\partial x}\bigg]_{(x_0, y_0, z_0)} dx$$

$$+ \frac{\partial g}{\partial y}\bigg]_{(x_0, y_0, z_0)} dy + \frac{\partial g}{\partial z}\bigg]_{(x_0, y_0, z_0)} dz$$

for dx, dy, and dz small.

Example 7.1 Let us estimate $(1.05)^2(2.99)^3$.

Let $f(x, y) = x^2 y^3$. We know that $f(1, 3) = 1^2 \cdot 3^3 = 27$. Let $dx = 0.05$ and $dy = -0.01$. Now

$$\frac{\partial f}{\partial x} = 2xy^3 \qquad \text{and} \qquad \frac{\partial f}{\partial y} = 3x^2 y^2.$$

Then

$$f(1.05, 2.99) = f(1 + dx, 3 + dy)$$

$$\approx 27 + \frac{\partial f}{\partial x}\bigg]_{(1,3)} (0.05) + \frac{\partial f}{\partial y}\bigg]_{(1,3)} (-0.01)$$

$$= 27 + 54(0.05) + 27(-0.01) = 27 + 2.70 - 0.27$$

$$= 29.43.$$

Actually, $(1.05)^2(2.99)^3 = 29.4708161475$. $\|$

7.2 Rate of Change

We turn now to rate of change questions. Reasoning precisely as in Section 2, we see that

$$\frac{\partial f}{\partial x}\bigg]_{(x_0, y_0)}$$

gives the rate that $f(x, y)$ changes per unit change in x at x_0 when y is held constant at y_0. A simple application of this observation is given in the following example.

Example 7.2 The volume V of a right circular cone of base radius r and height h is $\frac{1}{3}\pi r^2 h$. If the radius of the base increases at a constant rate of 4 in./min while the height remains constant at 12 in., let us find the rate of increase of the volume when $r = 6$ in.
    We have

$$\frac{\partial V}{\partial r} = \frac{1}{3}\pi(2r)h = \frac{2}{3}\pi r h.$$

When $r = 6$ and $h = 12$, we find that

$$\frac{\partial V}{\partial r} = \frac{2}{3}\pi \cdot 6 \cdot 12 = 48\pi.$$

Thus when $r = 6$, V is increasing 48π cu in. per inch change in r. If r is changing at the rate of 4 in./min, we see that the volume is changing at the rate of $(48\pi)4 = 192\pi$ in.^{3}/min. $\|$

    Rate of change problems are easier to handle if you have a chain rule for differentiation at your disposal. You may have recognized that a disguised chain rule argument was used in Example 7.2. Suppose that $z = f(x, y)$, $x = g(t)$, and $y = h(t)$ for some functions f, g, and h. Given a value of t, we can find the corresponding values of x and y, and therefore a value of z. Hence z appears as a function of t. The rate of change of z with respect to t, dz/dt, can be expressed in terms of the rates at which z changes with respect to x and y, and the rates at which x and y change with respect to t. This is surely not too surprising. The total rate of change of z is the sum of the rates of change due to the changing quantities x and y. These rates of change are, by the chain rule in Section 3,

$$\frac{\partial z}{\partial x}\frac{dx}{dt} \quad \text{and} \quad \frac{\partial z}{\partial y}\frac{dy}{dt}.$$

Although we do not claim to have proved it, the valid formula

$$\frac{dz}{dt} = \frac{\partial z}{\partial x}\frac{dx}{dt} + \frac{\partial z}{\partial y}\frac{dy}{dt}$$

should seem reasonable to you.

Example 7.3 Let us repeat Example 7.2 using a chain rule. We are given that

$$\frac{dr}{dt} = 4 \quad \text{and} \quad \frac{dh}{dt} = 0.$$

Now

$$\frac{dV}{dt} = \frac{\partial V}{\partial r}\frac{dr}{dt} + \frac{\partial V}{\partial h}\frac{dh}{dt} = \tfrac{2}{3}\pi rh\,\frac{dr}{dt} + \tfrac{1}{3}\pi r^2\,\frac{dh}{dt}.$$

Putting $dr/dt = 4$, $dh/dt = 0$, $r = 6$, and $h = 12$, we obtain

$$\frac{dV}{dt} = \tfrac{2}{3}\pi(6)(12)\cdot 4 + \tfrac{1}{3}\pi(6^2)\cdot 0 = 192\pi. \;\|$$

Analogous chain rules hold in situations involving more variables. For example, if $w = f(x, y, z)$, $x = g(t)$, $y = h(t)$, and $z = k(t)$, then

$$\frac{dw}{dt} = \frac{\partial w}{\partial x}\frac{dx}{dt} + \frac{\partial w}{\partial y}\frac{dy}{dt} + \frac{\partial w}{\partial z}\frac{dz}{dt}.$$

Partial derivatives can be computed similarly by a chain rule. For example, if $z = f(x, y)$, $x = g(u, v)$, and $y = h(u, v)$, then z appears as a function of u and v, and

$$\frac{\partial z}{\partial u} = \frac{\partial z}{\partial x}\frac{\partial x}{\partial u} + \frac{\partial z}{\partial y}\frac{\partial y}{\partial u}.$$

Such chain rules are valid for sufficiently "nice" functions; in particular, they hold for all rational functions.

Example 7.4 The volume V of a right circular cylinder of radius r and height h is given by $V = \pi r^2 h$. Suppose the volume is increasing at a rate of 72π cu in./min while the height is decreasing at a rate of 4 in./min. Let us find the rate of increase of the radius when the height is 3 in. and the radius is 6 in.

We have

$$\frac{dV}{dt} = \frac{\partial V}{\partial r}\frac{dr}{dt} + \frac{\partial V}{\partial h}\frac{dh}{dt},$$

so

$$\frac{dV}{dt} = 2\pi rh\,\frac{dr}{dt} + \pi r^2\,\frac{dh}{dt}.$$

We know that when $r = 6$ and $h = 3$,

$$\frac{dV}{dt} = 72\pi \quad \text{and} \quad \frac{dh}{dt} = -4.$$

(The negative sign occurs because h is *decreasing*.) We want to find dr/dt.

Substituting, we obtain

$$72\pi = 2\pi(6)(3)\frac{dr}{dt} + \pi(6^2)(-4),$$

so

$$36\pi\frac{dr}{dt} = 216\pi.$$

Hence

$$\frac{dr}{dt} = \frac{216\pi}{36\pi} = 6 \text{ in./min. } \parallel$$

EXERCISES

7.1 Estimate $\sqrt{(4.1)(.95)}$. [*Hint:* Approximate the function $\sqrt{xy}$ by a linear function at (4, 1).]

7.2 Estimate $(9.01)(3.98)^2 + (6.05)^2(3.98)$.

7.3 Given $z = x^2 - 2xy + xy^3$, $x = t^3 + 1$, and $y = 1/t$, find dz/dt when $t = 1$.

7.4 Given $w = x^2yz$, $x = 2t + 1$, $y = 3t^2$, and $z = t^2 - 2$, find dw/dt when $t = -1$.

7.5 Let $z = x^2 + 1/y^2$, $x = t^2$, and $y = t + 1$.

a) Find x, y, and z when $t = 1$.

b) Find

$$\left.\frac{dz}{dt}\right]_{t=1},$$

using the chain rule.

c) Express z as a function of t by substitution.

d) Find

$$\left.\frac{dz}{dt}\right]_{t=1}$$

by differentiating your answer to part (c). Compare with your answer in (b).

7.6 Given $w = x^2 + yz$, $x = uv$, $y = u - v$, and $z = 2u^2v$, find $\partial w/\partial u$ when $u = -1$ and $v = 2$.

7.7 A rectangular box is 14 in. wide, 20 in. long, and 8 in. high. Estimate the volume of material used in the construction of the box if the sides and bottom are $\frac{1}{8}$ in. thick and the box has no top.

7.8 A cylindrical silo with a hemispherical cap has volume $V = \pi r^2 h + \frac{2}{3}\pi r^3$, where h is the height of the cylinder and r is its radius. Estimate the change in volume of a silo with cylinder of 6-ft radius and 30-ft height if the radius is increased by 4 in. and the height h is decreased by 6 in.

7.9 By Newton's Law of Gravitation, the force of attraction between two bodies of masses m_1 and m_2 is

$$\frac{Gm_1m_2}{s^2},$$

where s is the distance between the bodies and G is the universal gravitational constant. Find the rate of change of the force of attraction between two bodies of constant masses of $(10)^4$ and $(10)^7$ units which are a distance of $(10)^4$ units apart and are separating at a rate of $(10)^2$ units distance per unit time. (Assume the given units are compatible, and don't worry about the value of G or the name of the units in the answer.)

7.10 Repeat Exercise 7.9 for the case in which the first body is gaining mass at the rate of 30 units mass per unit time, and the second body is losing mass at the rate of 80 units mass per unit time, while the other data remains the same.

REFERENCES

Practically all full-course calculus texts contain the material presented in this chapter.

6 | Integral Calculus

The importance of integral calculus, as well as differential calculus, lies in its usefulness in studying change. Indeed, differential and integral calculus are closely connected, as we shall show in Section 2. Newton and Leibniz were also responsible for the development of integral calculus in the form we know it today. However, the Greek mathematician Archimedes (287–212 B.C.) used the principles which lie at the heart of the subject in his work on the determination of the areas of certain types of planar regions. For this work alone, which was 2000 years ahead of its time, Archimedes must be regarded as one of the greatest mathematicians of recorded history. When Newton and Leibniz appeared on the scene, it was the natural time for calculus to be developed, as evidenced by their simultaneous but independent achievements in this field. Newton and Leibniz had the analytic geometry of the French mathematicians Pierre de Fermat (1601–1665) and René Descartes (1596–1650) on which to build; Archimedes did not.

We shall present the main ideas and a sample of applications of the integral calculus of one real variable. Integral calculus for functions of several real variables is a natural extension and is not difficult, but it is rather messy to write down in a book at this level, so we shall not attempt it.

1. THE DEFINITE INTEGRAL

1.1 Upper and Lower Sums

As is customary, we shall approach the definite integral by trying, like Archimedes, to compute the area of a region in the Euclidean plane. You know that the area of a rectangular region is the product of its length and width, and you can also find the areas of certain other familiar figures, such as a triangle, a trapezoid, a parallelogram, and a circle, although you have probably never seen a proof of the formula $A = \pi r^2$ for the area of a circle of radius r. However, if the boundary of a region is not made up of straight line segments or circular arcs, the problem is much harder.

Let us tackle the problem of finding the area of a region bounded on three sides by straight line segments and on the fourth side by the graph of a continuous function f mapping a closed interval $[a, b]$ into **R**; such a region is shaded in Fig. 1.1. We shall refer to this region as the *region from a to b under the graph of the function f*. In this section we shall give some methods of

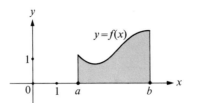

Fig. 1.1

estimating the area of such a region, and in the next section we shall show
how to compute the precise area.

One method of estimating this area is to use *upper sums* or *lower sums*.
Suppose $[a, b]$ is divided into n equal pieces by points $x_0, x_1, \ldots, x_n$, where
$x_0 = a$, $x_n = b$, and $x_i - x_{i-1} = (b - a)/n$, as illustrated in Fig. 1.2. Let
M_i be the maximum value of $f(x)$ for x in the interval $[x_{i-1}, x_i]$. Then

$$M_i(x_i - x_{i-1}) = M_i\left(\frac{b - a}{n}\right)$$

is the area of the shaded rectangle in Fig. 1.3, and is at least as large as the
area of the region under the graph of f from x_{i-1} to x_i, as we see from the
figure. Similarly, if m_i is the smallest value of $f(x)$ for x in $[x_{i-1}, x_i]$, then

$$m_i(x_i - x_{i-1}) = m_i\left(\frac{b - a}{n}\right)$$

gives the area of the shaded rectangle in Fig. 1.4, and is surely no larger than
the area of the region under the graph of f from x_{i-1} to x_i. If we introduce
the notation "$A_a^b(f)$" for the area of the region under the graph of f from a to

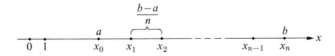

Fig. 1.2

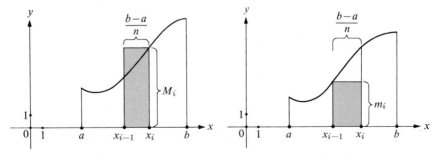

Fig. 1.3   Fig. 1.4

b, we then have

$$m_i\left(\frac{b-a}{n}\right) \leq A_{x_{i-1}}^{x_i}(f) \leq M_i\left(\frac{b-a}{n}\right).$$

Since obviously $A_a^b(f) = A_{x_0}^{x_1}(f) + A_{x_1}^{x_2}(f) + \cdots + A_{x_{n-1}}^{x_n}(f)$, we obtain

$$\left(\frac{b-a}{n}\right)(m_1 + m_2 + \cdots + m_n)$$
$$\leq A_a^b(f) \leq \left(\frac{b-a}{n}\right)(M_1 + M_2 + \cdots + M_n).$$

Definition. The *nth upper sum S_n for f from a to b* is

$$\left(\frac{b-a}{n}\right)(M_1 + M_2 + \cdots + M_n),$$

and the *nth lower sum s_n for f from a to b* is

$$\left(\frac{b-a}{n}\right)(m_1 + m_2 + \cdots + m_n).$$

For each $n \in \mathbf{Z}^+$, we have

$$s_n \leq A_a^b(f) \leq S_n.$$

Example 1.1 Let $f: \mathbf{R} \to \mathbf{R}$ be the function x^2 and consider the interval $[0, 1]$. For this interval and for $n = 2$, we have $x_0 = 0$, $x_1 = \frac{1}{2}$, and $x_2 = 1$, while

$$\frac{b-a}{n} = \frac{1-0}{2} = \frac{1}{2}.$$

It is easily seen from the graph of f in Fig. 1.5 that $m_1 = 0$ and $m_2 = \frac{1}{4}$, while $M_1 = \frac{1}{4}$ and $M_2 = 1$. Thus we have

$$s_2 = (\tfrac{1}{2})(0 + \tfrac{1}{4}) \leq A_0^1(x^2) \leq (\tfrac{1}{2})(\tfrac{1}{4} + 1) = S_2,$$

so

$$\tfrac{1}{8} \leq A_0^1(x^2) \leq \tfrac{5}{8}.$$

Of course, these bounds on $A_0^1(x^2)$ are very crude. ‖

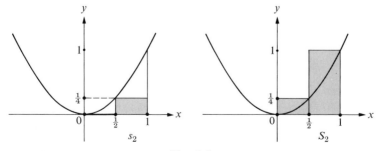

Fig. 1.5

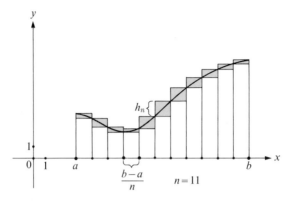

Fig. 1.6

1.2 The Definite Integral

The accuracy of our approximations s_n and S_n for $A_a^b(f)$ can be measured by $S_n - s_n$. Geometrically, $S_n - s_n$ corresponds to the sum of the areas of the little shaded rectangles along the graph in Fig. 1.6. Let h_n be the height of the largest of these rectangles; that is, h_n is the maximum of $M_i - m_i$. Then the sum of the areas of the little rectangles is less than $h_n(b - a)$, so

$$S_n - s_n \leq h_n(b - a).$$

For a continuous function f, it can be shown that as n gets large so that the horizontal dimension of the little shaded rectangles in Fig. 1.6 gets very small, the vertical dimensions get small also. That is, h_n is close to zero for sufficiently large n. We denote this by

$$\text{``} \lim_{n \to \infty} h_n = 0 \text{''}.$$

Thus, as n gets large, $S_n - s_n$ gets close to zero, and since $s_n \leq A_a^b(f) \leq S_n$, we see that both S_n and s_n must approach $A_a^b(f)$. We denote this by

$$\text{``} \lim_{n \to \infty} s_n = \lim_{n \to \infty} S_n = A_a^b(f) \text{''}.$$

This is a very important fact, and we state it as a theorem. While we have tried to make this theorem seem reasonable, we have not given a rigorous proof of it.

Theorem 1.1 *Let f be a continuous map of $[a, b]$ into $\mathbf{R}$. Then $\lim_{n \to \infty} S_n$ and $\lim_{n \to \infty} s_n$ both exist and are equal.*

Definition. The common value $A_a^b(f)$ of $\lim_{n \to \infty} S_n$ and $\lim_{n \to \infty} s_n$ as given by Theorem 1.1 is the **definite integral from a to b of the function f**, and is denoted by

$$\text{``} \int_a^b f \text{''}.$$

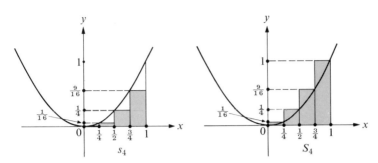

Fig. 1.7

Thus "$\int_a^b f$" is simply another notation for $A_a^b(f)$. The *integral sign* $\int$ is an elongated letter S, standing for *sum*.

According to Theorem 1.1, we might expect to improve our estimate of $A_0^1(x^2) = \int_0^1 x^2$ given in Example 1.1 by taking a larger value of n than 2.

Example 1.2 Let us estimate $\int_0^1 x^2$ by finding s_n and S_n with $n = 4$. As indicated in Fig. 1.7, we have $m_1 = 0$, $m_2 = \frac{1}{16}$. $m_3 = \frac{1}{4}$, and $m_4 = \frac{9}{16}$, while $M_1 = \frac{1}{16}$, $M_2 = \frac{1}{4}$, $M_3 = \frac{9}{16}$, and $M_4 = 1$. Of course,

$$\frac{b - a}{n} = \frac{1 - 0}{4} = \frac{1}{4}.$$

Thus

$$(\tfrac{1}{4})(0 + \tfrac{1}{16} + \tfrac{1}{4} + \tfrac{9}{16}) \leq \int_0^1 x^2 \leq (\tfrac{1}{4})(\tfrac{1}{16} + \tfrac{1}{4} + \tfrac{9}{16} + 1).$$

The arithmetic works out to give

$$\tfrac{7}{32} \leq \int_0^1 x^2 \leq \tfrac{15}{32}.$$

This is an improvement over Example 1.1 which gave us

$$\tfrac{4}{32} \leq \int_0^1 x^2 \leq \tfrac{20}{32}.$$

We might estimate $\int_0^1 x^2$ by averaging $\tfrac{7}{32}$ and $\tfrac{15}{32}$, arriving at $\tfrac{11}{32}$. We shall show in Section 2 that the exact value of $\int_0^1 x^2$ is $\tfrac{1}{3}$. Thus $\tfrac{11}{32}$ is not a bad estimate. ‖

We have drawn the graph of f in all our figures for the case in which $f(x) \geq 0$ for $a \leq x \leq b$. If $f(x) < 0$ for x in $[x_{i-1}, x_i]$, then both m_i and M_i are negative, so the contributions $[(b - a)/n]m_i$ to s_n and $[(b - a)/n]M_i$ to S_n are both negative. It is obvious that for functions f where $f(x)$ is sometimes negative and sometimes positive, $\int_a^b f$ equals the total area given by the portions of the graph above the x-axis minus the total area given by the portions of the graph below the x-axis, as illustrated in Fig. 1.8.

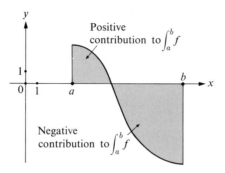

Fig. 1.8

1.3 Estimating by Trapezoids

Another useful device for estimating $\int_a^b f$ is to draw the chord of the graph connecting the points $(x_{i-1}, f(x_{i-1}))$ and $(x_i, f(x_i))$, as illustrated in Fig. 1.9. The interval $[x_{i-1}, x_i]$ then appears as the end of a trapezoid, shown shaded in Fig. 1.9, with its parallel sides vertical. Since the area of a trapezoid is equal to the product of half the sum of the lengths of its parallel sides (bases) and the distance between them (the altitude), we see that the area of the shaded trapezoid in Fig. 1.9 is

$$\frac{f(x_{i-1}) + f(x_i)}{2} \cdot \frac{b - a}{n}.$$

Example 1.3 Let us estimate $\int_0^1 x^2$ for $n = 2$, using trapezoids, as shown in Fig. 1.10. The area of the trapezoid over $[0, \frac{1}{2}]$ is

$$\frac{0 + \frac{1}{4}}{2} \cdot \frac{1}{2} = \frac{1}{16},$$

and the area of the trapezoid over $[\frac{1}{2}, 1]$ is

$$\frac{\frac{1}{4} + 1}{2} \cdot \frac{1}{2} = \frac{5}{16}.$$

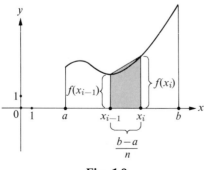

Fig. 1.9

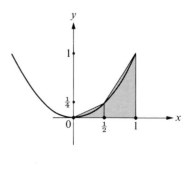

Fig. 1.10

Thus we estimate that $\int_0^1 x^2$ is about $\frac{1}{16} + \frac{5}{16} = \frac{6}{16} = \frac{3}{8}$. As we stated in Example 1.2, the actual value of $\int_0^1 x^2$ is $\frac{1}{3}$. ‖

EXERCISES

1.1 Find the upper sum S_2 and the lower sum s_2 for x^2 from 0 to 2. (The actual value of $\int_0^2 x^2$ is $\frac{8}{3}$.)

1.2 Find the upper sum S_4 and the lower sum s_4 for x^2 from 0 to 2. Compare with your answer in Exercise 1.1.

1.3 Let $f(x) = 1 - x$ and consider the interval $[0, 2]$.

a) Sketch the graph of f over this interval.
b) Find s_2 for this interval.
c) Find S_2 for this interval.
d) From the graph of f in part (a), give the exact value of $\int_0^2 f$.

1.4 a) Find the upper sum S_4 and the lower sum s_4 for x^3 from -1 to 1.
b) From the graph of x^3, give the exact value of $\int_{-1}^1 x^3$.

1.5 Find S_4 and s_4 for the function $1/x$ from 1 to 2.

1.6 For $a, b \in \mathbf{R}^+$, where $a < b$, how does $\int_a^b x^n$ compare with $\int_{-b}^{-a} x^n$ for $n \in \mathbf{Z}$? Draw figures and consider the following cases.

a) $n = 0$ b) $n > 0$ and even   c) $n > 0$ and odd
d) $n < 0$ and even   e) $n < 0$ and odd

1.7 Give the exact value of $\int_{-2}^3 4$, where $4: \mathbf{R} \to \mathbf{R}$ is the usual constant function.

1.8 Argue that for a function f whose graph over $[a, b]$ is a straight line, the estimate of $\int_a^b f$ given by trapezoids as in Example 1.3 is exact.

1.9 Let $[a, b]$ be divided into n equal pieces by points $x_0 = a, x_1, \ldots, x_n = b$. Let f be a continuous function with domain $[a, b]$, and let $y_i = f(x_i)$. Show that the estimate of $\int_a^b f$ given by trapezoids as in Example 1.3 is

$$\left(\frac{b - a}{2n} \right)(y_0 + 2y_1 + \cdots + 2y_{n-1} + y_n).$$

(This formula is the *trapezoidal rule*.)

1.10 Estimate $\int_0^1 x^2$ by the trapezoidal rule (see Exercise 1.9) for $n = 4$, and compare this estimate with the values of S_4 and s_4 found in Example 1.2.

1.11 Sketch the graph of a function f over the interval $[0, 2]$ for which S_2 is much closer to $\int_0^2 f$ than the average of S_2 and s_2. (This shows that an averaging technique such as the one used in Example 1.2 may not give good results.)

1.12 Sketch the graph of a function f over the interval $[0, 6]$ for which $S_3 > S_2$. (This shows that while for large n, S_n approaches $\int_a^b f$, we need not have $S_{n+1} \leq S_n$.)

1.13 Show that for any function f mapping $[a, b]$ into $\mathbf{R}$, $S_{2n} \leq S_n$ for each n. Compare with Exercise 1.12.

2. THE FUNDAMENTAL THEOREM

2.1 The Fundamental Theorem

In the preceding section we defined $\int_a^b f$ and had some practice in estimating this definite integral. In this section we shall indicate how the precise value of $\int_b^a f$ can often be easily computed.

Let f be a continuous function mapping $[a, b]$ into $\mathbf{R}$. Let F be the function mapping $[a, b]$ into $\mathbf{R}$ defined by

$$F(x) = \int_a^x f$$

for x in $[a, b]$. For each x in $[a, b]$, the value of $F(x)$ is given geometrically by the area of the region under the graph of f from a to x, shown shaded in Fig. 2.1. Note in particular that $F(a) = \int_a^a f = 0$.

The trick is to differentiate F. By definition,

$$F'(x_0) = \lim_{dx \to 0} \frac{F(x_0 + dx) - F(x_0)}{dx} = \lim_{dx \to 0} \frac{\int_a^{x_0+dx} f - \int_a^{x_0} f}{dx}.$$

Since $\int_a^{x_0+dx} f$ is the area of the region under the graph of f from a to $x_0 + dx$, and $\int_a^{x_0} f$ is the area of the region under the graph from a to x_0, the difference $\int_a^{x_0+dx} f - \int_a^{x_0} f$ gives the area of the region under the graph of f between x_0 and $x_0 + dx$, shown shaded in Fig. 2.2. That is,

$$\int_a^{x_0+dx} f - \int_a^{x_0} f = \int_{x_0}^{x_0+dx} f.$$

Let M_{dx} be the largest value of $f(x)$ for x in the interval $[x_0, x_0 + dx]$, and let m_{dx} be the smallest value of $f(x)$ over this interval. Then, as we saw in the preceding section,

$$(dx)(m_{dx}) \leq \int_{x_0}^{x_0+dx} f \leq (dx)(M_{dx}).$$

Thus

$$(dx)(m_{dx}) \leq \int_a^{x_0+dx} f - \int_a^{x_0} f \leq (dx)(M_{dx}),$$

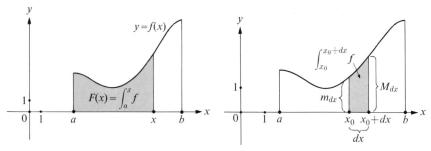

Fig. 2.1 Fig. 2.2

so

$$m_{dx} \leq \frac{F(x_0 + dx) - F(x_0)}{dx} \leq M_{dx}.$$

Since f is continuous, as dx approaches 0, both m_{dx} and M_{dx} approach $f(x_0)$. Since

$$\frac{F(x_0 + dx) - F(x_0)}{dx}$$

is caught between m_{dx} and M_{dx}, we must have

$$F'(x_0) = \lim_{dx \to 0} \frac{F(x_0 + dx) - F(x_0)}{dx} = f(x_0).$$

Thus the derivative of F, the area function given by $F(x) = \int_a^x f$, exists and is the function f. We have demonstrated the main theorem of this chapter.

Theorem 2.1 *Let f be a continuous function mapping $[a, b]$ into $\mathbf{R}$. Let F mapping $[a, b]$ into $\mathbf{R}$ be defined by*

$$F(x) = \int_a^x f$$

for x in $[a, b]$. Then F' exists for x in $]a, b[$, and we have $F'(x) = f(x)$.

Corollary. *Let f be continuous in an open interval containing $[a, b]$ and let G be any antiderivative of f in $[a, b]$. Then*

$$\int_a^b f = G(b) - G(a).$$

Proof. Let F mapping $[a, b]$ into $\mathbf{R}$ be defined as in Theorem 2.1. Since

$$\frac{dF}{dx} = \frac{dG}{dx} = f,$$

we have

$$\frac{d(F - G)}{dx} = \frac{dF}{dx} - \frac{dG}{dx} = f - f = 0,$$

the constant function 0. We saw in the preceding chapter that the only functions with derivative zero are the constant functions. Thus for $x \in [a, b]$, we have

$$F(x) - G(x) = c$$

for some $c \in \mathbf{R}$. Recall that

$$F(a) = \int_a^a f = 0.$$

Thus, putting $x = a$, we obtain

$$F(a) - G(a) = 0 - G(a) = c,$$

so $c = -G(a)$, and $F(x) - G(x) = -G(a)$. Putting $x = b$, we obtain

$$F(b) - G(b) = -G(a),$$

so

$$F(b) = \int_a^b f = G(b) - G(a). \ \blacksquare$$

Theorem 2.1, together with its corollary, is known as *The Fundamental Theorem of Integral Calculus*.

2.2 Computing $\int_a^b f$

The corollary of Theorem 2.1 gives us a powerful method for evaluating $\int_a^b f$
Recall that an antiderivative of the function cx^n is

$$c\,\frac{x^{n+1}}{n+1} \qquad \text{if} \qquad n \neq -1.$$

Example 2.1 Let us find the exact value of $\int_0^1 x^2$, which we estimated several times in the preceding section. An antiderivative G of the function x^2 is $x^3/3$. Thus by our corollary, we have

$$\int_0^1 x^2 = G(1) - G(0) = \tfrac{1}{3} - \tfrac{0}{3} = \tfrac{1}{3}.$$

After all our work trying to get a good estimate for this definite integral in the preceding section, you should appreciate the elegance and beauty of this easy computation. ‖

It is customary to denote $G(b) - G(a)$ by "$G(x)]_a^b$". The **upper limit of integration** is b and the **lower limit** is a. For example, you will usually see $\int_0^1 x^2$ computed as follows:

$$\int_0^1 x^2 = \frac{x^3}{3}\bigg]_0^1 = \frac{1}{3} - \frac{0}{3} = \frac{1}{3}.$$

To compute a definite integral as illustrated above, we have to be able to find an antiderivative of the function f. As we saw in the last chapter, if F is any antiderivative of f, then every antiderivative is of the form $F + c$ for some constant function c. Thus $F + c$, where c is an "arbitrary constant function," is the most general antiderivative of f. These antiderivatives play a crucial role in the evaluation of definite integrals, and the general antiderivative $F + c$ is known as the **indefinite integral of** f and is denoted by "$\int f$". A definite integral has limits a and b and is a *number*, while an indefinite integral has no given limits and is a *collection of functions*.

Example 2.2 The indefinite integral $\int (x^3 + 2x^2 + 1)$ is

$$\frac{x^4}{4} + 2\frac{x^3}{3} + x + c,$$

where c is an arbitrary constant function. ‖

Tables have been prepared which give the indefinite integrals of many functions. The best way to find an indefinite integral of a function f is to look it up in a table, unless f is a very simple function. We conclude with one more easy example.

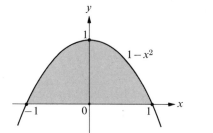

Fig. 2.3

Example 2.3 Let us sketch the region bounded by the graph of the polynomial function $1 - x^2$ and the x-axis, and let us find the area of the region. The function $1 - x^2$ obviously has a maximum at $x = 0$ and $f(x)$ decreases as x gets further from 0. The graph crosses the x-axis when $1 - x^2 = 0$, that is, when $x = \pm 1$. The graph is shown in Fig. 2.3, where we have shaded the region whose area we wish to find. Clearly, the area is $\int_{-1}^{1} (1 - x^2)$ square units, and we have

$$\int_{-1}^{1} (1 - x^2) = \left(x - \frac{x^3}{3} \right)\bigg]_{-1}^{1} = \left(1 - \frac{1}{3} \right) - \left(-1 - \frac{-1}{3} \right)$$

$$= \frac{2}{3} - \left(-\frac{2}{3} \right) = \frac{4}{3}. \; ‖$$

EXERCISES

2.1 Find each of the following indefinite integrals.

a) $\int 2$ b) $\int 0$ c) $\int x$ d) $\int (3x^2 + 1)$

e) $\int (x^4 + 3x^2 - 5x + 7)$ f) $\int \sqrt{x}$

2.2 Find each of the following indefinite integrals.

a) $\int \frac{1}{x^2}$ b) $\int \left(3x^2 + \frac{1}{x^2} \right)$ c) $\int \left(x^5 - \frac{3}{x^3} \right)$

d) $\displaystyle\int\left(\frac{x^2-3x}{x^4}\right)$ 　　 e) $\displaystyle\int\left(\sqrt{x}+\frac{1}{\sqrt{x}}\right)$ 　　 f) $\displaystyle\int\left(\frac{x^{3/2}-x}{\sqrt{x}}\right)$

2.3 Evaluate each of the following definite integrals.

a) $\displaystyle\int_0^1 (x^3+3x^2)$ 　　 b) $\displaystyle\int_{-1}^2 (x+1)$ 　　 c) $\displaystyle\int_{-2}^2 (x^3-3x)$

d) $\displaystyle\int_1^2\left(x^4-3x^2+\frac{2}{x^2}\right)$ 　　 e) $\displaystyle\int_0^2 \sqrt{2x}$

2.4 Sketch the graph of the function $4-x^2$ and find the area of the region bounded by the graph and the x-axis.

2.5 Sketch the graph of the function x^4-1 and find the area of the region bounded by the graph and the x-axis.

2.6 Sketch the graph of the function x^2-4x and find the area of the region bounded by the graph and the x-axis.

2.7 Show that $\int_{-a}^a x^n = 0$ if n is an odd positive integer, i.e., if $n = 1, 3, 5,$ etc.

2.8 Show that $\int_{-a}^a f = 0$ if f is a polynomial function containing only odd positive powers of the variable x. [*Hint:* See Exercise 2.7.]

2.9 Let f and g be continuous maps of **R** into **R**. Argue geometrically, using areas, that each of the following relations is valid for all $a, b, c \in$ **R**, where $a < b < c$.

a) $\displaystyle\int_a^b f + \int_b^c f = \int_a^c f$ 　　 b) $\displaystyle\int_a^b (-f) = -\int_a^b f$

c) $\displaystyle\int_a^b (f+g) = \int_a^b f + \int_a^b g$ 　　 d) $\displaystyle\int_a^b (cf) = c\int_a^b f$

2.10 Argue that each of the relations in Exercise 2.9 is valid by using the corollary of Theorem 2.1.

2.11 a) Show that if the corollary of Theorem 2.1 is to hold for all $a, b \in$ **R**, where a may be either greater than or less than b, we must define $\int_a^b f = -\int_b^a f$ for $a > b$.

b) Show that we must define $\int_a^b f = -\int_b^a f$ for $a > b$ if we wish the relation in Exercise 2.9(a) to hold for all $a, b, c \in$ **R**. [*Hint:* Let $c = a$.]

3. THE NATURE OF APPLICATIONS

3.1 Integrals and Products

We have seen that for a continuous map f of $[a, b]$ into **R**, the definite integral $\int_a^b f$ gives the area of the region under the graph of f from a to b, shown shaded in Fig. 3.1. An area of a plane configuration is roughly the product of its dimensions. Just as with differential calculus, the power of integral calculus lies in its ability to handle situations in which quantities vary. For example, the area of a rectangle is the product of its length and its height, and for a rectangle, these quantities remain constant throughout the region. For the

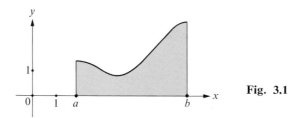

Fig. 3.1

shaded region in Fig. 3.1, however, the height varies as you travel across the region from left to right. This is the type of situation in which calculus comes into play.

The definite integral is useful in any type of application in which, in the case of quantities which do not vary, you would compute a *product*. (Since any quantity can be viewed as the product of itself and 1, we are talking about a very general situation.) For example, if the length and height of a plane region do not vary over the region, the area of the region is the product of these dimensions. We list here several more quantities which appear as a product of two other quantities.

Volume. For a solid with cross section of constant area A and of constant height h, the volume V is the product Ah.

Work. If a body is moved a distance s by means of a constant force F, the work W done in moving the body is the product Fs.

Distance. If a body travels for time t with a constant speed v, the distance s traveled is the product vt.

Speed. If a body travels for time t with a constant acceleration a, the speed v attained is the product at.

Force. If the pressure per square unit on a plane region of area A is a constant p throughout the region, then the total force F on the region due to this pressure is the product pA.

Moments. The moment M about an axis of a body of mass m, all points of which are a constant (*signed*) distance s from the axis, is the product ms.

Moments of inertia. The moment of inertia I about an axis of a body of mass m, all points of which are a constant distance s from the axis, is the product ms^2.

Some of these applications are discussed in the sections which follow.

3.2 A Useful Notation for Applications

In applications of the definite integral in which a product appears, it is often easier to use a slightly different notation than "$\int_a^b f$". Recall that $\int_a^b$ is the limit as n gets large of the upper sums S_n. Each summand of S_n can be

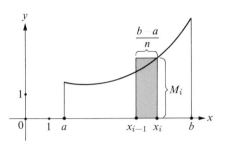

Fig. 3.2

regarded as the area of a rectangle; the rectangle corresponding to the summand $M_i[(b - a)/n]$ is shown shaded in Fig. 3.2. Now $M_i[(b - a)/n]$ is a *product*. Let us take local coordinates at the point $(x_{i-1}, 0)$, and let $dx = (b - a)/n$. Then

$$M_i\left(\frac{b - a}{n}\right) = M_i \, dx.$$

Similarly, as a contribution to the lower sum s_n, we have

$$m_i\left(\frac{b - a}{n}\right) = m_i \, dx.$$

If x is any point in the interval $[x_{i-1}, x_i]$, we have $m_i \leq f(x) \leq M_i$, so

$$m_i \, dx \leq f(x) \, dx \leq M_i \, dx.$$

We can think of $f(x) \, dx$ as giving the area of a rectangle of height $f(x)$, as shown in Fig. 3.3. Since

$$\lim_{n \to \infty} S_n = \lim_{n \to \infty} s_n = \int_a^b f,$$

and since $f(x) \, dx$ is caught between the contributions $m_i \, dx$ for s_n and $M_i \, dx$ for S_n, surely as n gets large, the sum of all the $f(x) \, dx$ for each interval $[x_{i-1}, x_i]$ also must approach $\int_a^b f$. For this reason, one often denotes $\int_a^b f$

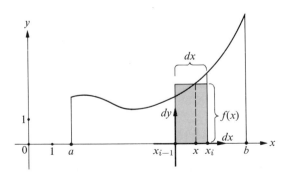

Fig. 3.3

by "$\int_a^b f(x)\,dx$". This notation is useful in applications, for you can think of $\int$ as meaning "Take the limit of the sum of all", and $f(x)\,dx$ appears as a *product*. Thus one takes the limit of the sum of all products $f(x)\,dx$, as the size of dx approaches zero. The sections which follow discuss some of the applications listed above, and illustrate how the notation "$\int_a^b f(x)\,dx$" is useful in forming the correct definite integral.

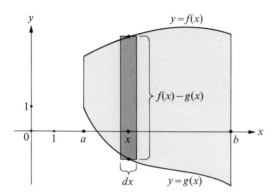

Fig. 4.1

4. THE AREA OF A PLANE REGION

In this section we shall show how to find the area of a more general plane region than one under the graph of a continuous function from a to b. Consider, for example, the shaded region in Fig. 4.1, which lies between the graphs of continuous functions f and g from a to b. We can estimate the area of this region by taking thin rectangles like the one shown heavily shaded in Fig. 4.1, and adding up the areas of such rectangles over the region. This rectangle is dx units wide and $f(x) - g(x)$ units high, for x as shown in Fig. 4.1. (Note that $g(x)$ is itself negative. The distance between the graphs of f and g over a point x is always $f(x) - g(x)$ if the graph of f is above that of g at x.) Thus the area of this rectangle is

$$(f(x) - g(x))\,dx.$$

We wish to add up the areas of such rectangles and take the limit of the resulting sum as dx grows smaller and the number of rectangles increases. As we saw in the preceding section, the limit of such a sum will be

$$\int_a^b (f(x) - g(x))\,dx.$$

We suggest that you use the following steps to find the area of a plane region.

STEP 1. Sketch the region, finding the points of intersection of bounding curves.

STEP 2. On your sketch, draw a typical thin rectangle with its small dimension either parallel to the x-axis and of length dx, or parallel to the y-axis and of length dy.

STEP 3. Looking at your sketch, write down the area of this rectangle as a product of the short dimension (dx or dy) and the long dimension.

STEP 4. Add up the areas found in Step 3 and take the limit of the resulting sum by integrating between the appropriate (x or y) limits.

We illustrate this procedure with two examples.

Example 4.1 Let us find the area of the plane region bounded by the curves with equations $y = x^2$ and $y = 3 - 2x$.

STEP 1. The curves are sketched and the region whose area is desired is shaded in Fig. 4.2. The points $(-3, 9)$ and $(1, 1)$ of intersection of the boundary curves are found by solving the equations $y = x^2$ and $y = 3 - 2x$ simultaneously.

STEP 2. A typical thin vertical rectangle is drawn and heavily shaded in Fig. 4.2.

STEP 3. The graph of the function $3 - 2x$ is above the graph of x^2 over this region, so the height of the thin rectangle over a point x is $(3 - 2x) - x^2$. The area of this thin rectangle is therefore $[(3 - 2x) - x^2]\, dx$.

STEP 4. Since we wish to add up our thin rectangles in the x-direction from $x = -3$ to $x = 1$, the appropriate integral is

$$\int_{-3}^{1} [(3 - 2x) - x^2]\, dx = \left(3x - 2\frac{x^2}{2} - \frac{x^3}{3}\right)\Bigg]_{-3}^{1}$$

$$= (3 - 1 - \tfrac{1}{3}) - (-9 - 9 + 9) = \tfrac{32}{3}. \;\|$$

Example 4.2 Let us try to solve the same problem that we solved in Example 4.1 by taking thin horizontal rectangles.

STEP 1. The appropriate sketch is given in Fig. 4.3.

STEP 2. Note that we can't take just one rectangle as typical for the whole region; for if the rectangle lies above the line given by $y = 1$, it is bounded on the right by the line with equation $y = 3 - 2x$, while a rectangle below the line $y = 1$ is bounded on both ends by the curve given by $y = x^2$. We split the region into two parts by the line with equation $y = 1$, and find the area of each part separately.

STEP 3. To find the horizontal dimensions of our rectangles, we must solve for x in terms of y, obtaining $x = \tfrac{1}{2}(3 - y)$ for the line and $x = \pm\sqrt{y}$ for the curve. The upper rectangle has an area of $[\tfrac{1}{2}(3 - y) - (-\sqrt{y})]\, dy$, while the lower rectangle has an area of $[\sqrt{y} - (-\sqrt{y})]\, dy$.

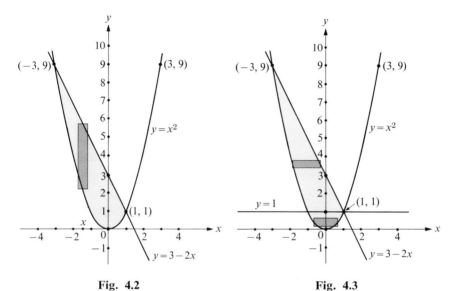

Fig. 4.2   Fig. 4.3

STEP 4. The appropriate integrals are

$$\int_1^9 [\tfrac{1}{2}(3 - y) - (-\sqrt{y})] \, dy = \left(\frac{3}{2} y - \frac{y^2}{4} + \frac{2}{3} y^{3/2} \right) \Bigg]_1^9$$

$$= (\tfrac{27}{2} - \tfrac{81}{4} + \tfrac{2}{3} \cdot 27) - (\tfrac{3}{2} - \tfrac{1}{4} + \tfrac{2}{3}) = \tfrac{28}{3}$$

and

$$\int_0^1 [\sqrt{y} - (-\sqrt{y})] \, dy = \int_0^1 2\sqrt{y} \, dy = 2 \cdot \tfrac{2}{3} y^{3/2} \Bigg]_0^1 = \tfrac{4}{3} - 0 = \tfrac{4}{3}.$$

Thus the total area is $\tfrac{28}{3} + \tfrac{4}{3} = \tfrac{32}{3}$ square units. Clearly, this computation is not as nice as that in Example 4.1. ‖

We call your attention again to the elegance and power of these techniques of calculus, as illustrated by these impressive applications.

EXERCISES

4.1 Find the area of the region bounded by the graphs of the function x^2 and the constant function 4.

4.2 Find the area of the region bounded by the graphs of the function x^4 and the constant function 1.

4.3 Find the area of the region bounded by the graphs of the functions x and x^2. [*Hint:* Draw a careful sketch.]

4.4 Find the total area of the regions bounded by the graphs of the functions x and x^3. [*Hint:* Draw a careful sketch.]

4.5 Find the total area of the regions bounded by the graphs of the functions x^4 and x^2. [*Hint:* Draw a careful sketch.]

4.6 Find the area of the region bounded by the graphs of the functions x^2 and x^3. [*Hint:* Draw a careful sketch.]

4.7 Find the area of the region bounded by the graphs of $x^4 - 1$ and $1 - x^2$.

4.8 Find the area of the region bounded by the graphs of $x^2 - 1$ and $x + 1$.

4.9 Find the area of the region bounded by the curves with equations $x = y^2$ and $y = x - 2$.

4.10 Find the area of the region bounded by the graphs of $\sqrt{x}$ and x^2.

5. VOLUMES OF SOLIDS OF REVOLUTION

As we observed in Section 3, the volume of a solid of constant height h and with cross sections of constant area A is the *product Ah*. This allows us to use integrals to compute volumes of solids for which neither the area of the cross sections nor the height remains constant. We illustrate with two examples. The steps in our computations are analogous to those we used to find areas in Section 4.

Example 5.1 Let the first-quadrant region bounded by the curves with equations $y = x^2$, $x = 0$, and $y = 1$ be revolved about the y-axis, and let us find the volume of the resulting solid.

STEP 1. The region is shown shaded in Fig. 5.1. Revolving the region about the y-axis gives a solid bowl, as shown in Fig. 5.2.

STEP 2. When the heavily shaded rectangle in Fig. 5.1 is revolved about the y-axis, it sweeps out a thin circular slab, as shown in Fig. 5.3.

STEP 3. The volume of such a slab is the product of the area of the circular face and the thickness (or height) of the slab. The area of a face is πx^2, for the point (x, y) shown in Fig. 5.3. The thickness of the slab is dy, so the volume of this slab is $\pi x^2\, dy$.

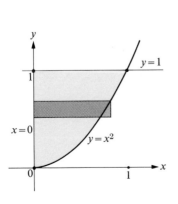

Fig. 5.1

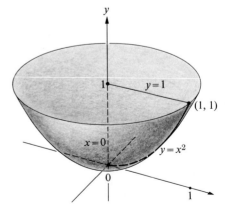

Fig 5.2

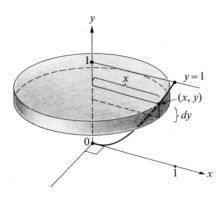

Fig. 5.3   Fig. 5.4

STEP 4. We wish to add up the volumes of such slabs from $y = 0$ to $y = 1$, and take the limit of the resulting sum as dy approaches 0. Since our limits are y-limits, we must express the volume $\pi x^2 \, dy$ entirely in terms of y. For the point (x, y) in Fig. 5.3, we have $x^2 = y$, so the volume of the slab becomes $(\pi y) \, dy$. Thus the appropriate integral is

$$\int_0^1 \pi y \, dy = \pi \frac{y^2}{2} \bigg]_0^1 = \frac{\pi}{2} - 0 = \frac{\pi}{2}. \; \|$$

Example 5.2 Let us repeat Example 5.1, but let us calculate the volume by taking thin vertical rectangles as shown in Fig. 5.4 (Steps 1 and 2).

STEP 3. As the heavily shaded rectangle in Fig. 5.4 is revolved about the y-axis, it sweeps out the cylindrical shell shown in Fig. 5.5. The volume of such a shell is approximately the product of the surface area of the cylinder and the thickness of the shell (the wall of the cylinder). The surface area in

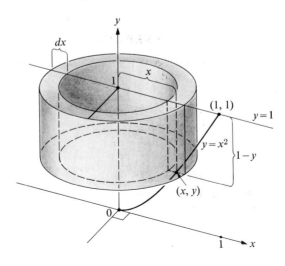

Fig. 5.5

turn is the product of the perimeter of the circle and the height of the cylinder. For the point (x, y) shown in Fig. 5.5, the perimeter is $2\pi x$ and the height is $1 - y$. Thus the volume of the cylindrical shell is $2\pi x(1 - y)\,dx$.

STEP 4. We want to add up the volumes of such cylindrical shells from $x = 0$ to $x = 1$, and take the limit of the resulting sum as dx approaches 0. With x-limits, we must express the volume of the shell entirely in terms of x. For the point (x, y) of Fig. 5.5, we have $y = x^2$, so the volume of the shell is $2\pi x(1 - x^2)\,dx = (2\pi x - 2\pi x^3)\,dx$. Thus the appropriate integral is

$$\int_0^1 (2\pi x - 2\pi x^3)\,dx = \left(2\pi \frac{x^2}{2} - 2\pi \frac{x^4}{4}\right)\Big]_0^1 = \left(\frac{2\pi}{2} - \frac{2\pi}{4}\right) - 0 = \frac{\pi}{2}. \;\|$$

The preceding examples illustrate the two basic methods of finding the volume of a solid of revolution: the *slab method* and the *shell method*. In working the exercises, be sure that you *draw a figure, and from your figure, write down the appropriate volume of your slab or shell.* In our examples we generated the solid by revolving a region about an axis forming a boundary of the region. Suppose a region is revolved about an axis not touching the region. For the slab method, in place of a circular slab you will get an *annular slab* (a slab with a hole in it), as shown in Fig. 5.6. The volume of such an annular slab is the volume of the solid disk slab minus the volume of the hole. Thus for r_1 and r_2 as shown in Fig. 5.6, the volume of such a slab of thickness dy is $(\pi r_1^2 - \pi r_2^2)\,dy$.

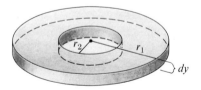

Fig. 5.6

EXERCISES

5.1 Find the volume of the solid generated by revolving the region bounded by the curves with equations $y = x^2$ and $y = 1$ about the line with equation $y = 1$, using circular slabs.

5.2 Repeat Exercise 5.1 using cylindrical shells.

5.3 Find the volume of the solid generated by revolving the region bounded by the curves given by $y = x^2$ and $y = 1$ about the line with equation $x = 2$, using the slab method.

5.4 Repeat Exercise 5.3 using cylindrical shells.

5.5 Verify the formula $V = \frac{1}{3}\pi r^2 h$ for the volume of a right circular cone of height h with base of radius r. [*Hint:* Rotate the region bounded by the lines with equations $y = (r/h)x$, $y = 0$, and $x = h$ about the x-axis.]

5.6 Verify the formula $V = \frac{4}{3}\pi r^3$ for the volume of a solid sphere (3-ball) of radius *r*. [*Hint:* Rotate the region above the *x*-axis bounded by the circle with equation $x^2 + y^2 = r^2$ and the line with equation $y = 0$ about the *x*-axis.]

5.7 Find the volume of the solid generated by rotating the region bounded by the graphs of the functions x and x^2 about the *y*-axis.

6. MISCELLANEOUS APPLICATIONS

In this section we shall present three examples illustrating three of the applications of the definite integral mentioned in Section 3. Additional applications are left to the student in the exercises.

Example 6.1 Given that the velocity of a body traveling on a number line at time t is $2t + t^2$, let us find how far the body travels from time $t = 2$ to time $t = 4$, using integral calculus. (We won't bother with the measurement units.) Since distance equals the product of speed and time, assuming that speed and time are constant, we see that at time t the distance traveled over the next small time interval dt is about $(2t + t^2)\, dt$ units. We wish to add up these distances as t ranges from 2 to 4, and take the limit of the resulting sum as dt approaches 0. The appropriate integral is

$$\int_2^4 (2t + t^2)\, dt = \left(2\frac{t^2}{2} + \frac{t^3}{3} \right)\Bigg]_2^4 = (\tfrac{32}{2} + \tfrac{64}{3}) - (\tfrac{8}{2} + \tfrac{8}{3})$$

$$= \tfrac{24}{2} + \tfrac{56}{3}$$

$$= 12 + \tfrac{56}{3}$$

$$= \tfrac{92}{3}. \parallel$$

In finding distance by taking a definite integral of the velocity, as in the preceding example, you must be sure that the velocity is always positive. The velocity of a body moving on the *x*-axis is considered positive if it is moving to the right and negative if the body is moving to the left. Thus the integral of the velocity from t_1 to t_2 will give the total distance traveled toward the right minus the distance traveled toward the left between these times. This "resultant distance" may not give a true picture of how far the body has traveled.

Example 6.2 The force F required to stretch a coil spring is proportional to the distance x it is stretched from its natural length. That is, $F = kx$ for some constant k, the *spring constant*. Suppose a spring is such that the force required to stretch it 1 ft from its natural length is 4 lbs. For this spring, $k = 4$. Let us find the work done in stretching the spring a distance of 4 ft from its natural length.

As we stated in Section 3, work is defined as the product of force and distance, if these two quantities are constant. Thus as our spring is stretched an additional small distance dx at a distance x from its natural length, the work done is approximately $F\,dx = 4x\,dx$. We wish to add all these little pieces of work from $x = 0$ to $x = 4$, and take the limit of the resulting sum as dx approaches 0. The appropriate integral is

$$\int_0^4 4x\,dx = 4\frac{x^2}{2}\bigg]_0^4 = \tfrac{64}{2} - 0 = 32 \text{ ft-lb (foot-pounds).} \|$$

Example 6.3 The fluid pressure per square foot at a depth of s ft in water is about $(62.4)s$ lbs. Suppose a dam 16 ft high has the shape of the region bounded by the curves with equations $y = x^2$ and $y = 16$. Let us find the total force due to water pressure on the dam when the water level is at the top of the dam.

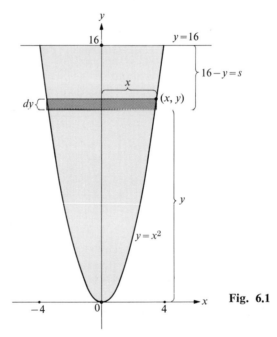

Fig. 6.1

The region bounded by the curves with equations $y = x^2$ and $y = 16$ is shaded in Fig. 6.1. For small dy, the pressure at a depth s on the heavily shaded strip in Fig. 6.1 is nearly constant at $(62.4)s$ lbs/ft². The area of this strip is $2x\,dy$, so the force on it is approximately $62.4s(2x\,dy)$ lbs. We want to add up these quantities $62.4s(2x\,dy)$ over the region as y ranges from 0 to 16, and take the limit of the resulting sum as dy approaches 0. We must express this quantity entirely in terms of y. For the point (x, y) in Fig. 6.1, we have $s = 16 - y$ and $y = x^2$, so $x = \sqrt{y}$. Thus

$$62.4s(2x\,dy) = 62.4(16 - y)2y^{1/2}\,dy = 124.8(16y^{1/2} - y^{3/2})\,dy.$$

The appropriate integral is

$$\int_0^{16} 124.8(16y^{1/2} - y^{3/2})\, dy = 124.8\left(16(\tfrac{2}{3})y^{3/2} - \tfrac{2}{5}y^{5/2}\right)\Big]_0^{16}$$

$$= 124.8\left(16(\tfrac{2}{3})64 - (\tfrac{2}{5})1024\right) - 0$$

$$= 124.8(1024)(\tfrac{2}{3} - \tfrac{2}{5})$$

$$= 124.8(1024)\tfrac{4}{15} = 34{,}078.72 \text{ lbs. } \|$$

EXERCISES

6.1 Find the work done in stretching the spring of Example 6.2 from 2 ft longer than its natural length to 6 ft longer than its natural length.

6.2 Given that the acceleration of a body starting from rest at time 0 is $3t$ at time t, find the velocity of the body and the distance the body has traveled since $t = 0$ at time $t = 4$. (Assume the units given are compatible.)

6.3 A thin metal plate is cut from a thin sheet of metal of mass 3 per square unit. Given that the plate just covers the first-quadrant region bounded by the curves with equations $y = x^2$, $y = 1$, and $x = 0$, find the moment of the plate about the y-axis. (Refer to Section 3 for the definition of the moment about an axis. Assume the units are compatible, and don't worry about the units for the answer.)

6.4 For the plate of Exercise 6.3, find its moment about the axis with equation $x = 2$.

6.5 For the plate in Exercise 6.3, find its moment of inertia about the x-axis. (Refer to Section 3 for the definition of the moment of inertia about an axis.)

6.6 A dam is in the shape of semicircle of radius 20 ft with the diameter of the circle at the top of the dam. Express the force on the dam when the water is at the top as a definite integral.

6.7 A cylindrical tank of radius 2 ft and height 10 ft is full of water and is emptied by dropping in a hose, pumping all the water up to the top of the tank, and letting it spill over. Find the work done. (A cubic foot of water weighs about 62.4 lbs.)

6.8 A particle of mass m at the point 10 on the x-axis is attracted toward the origin by a force of magnitude k/x^2. Find the work done by the force in moving the particle from 10 to 5.

REFERENCES

Almost any full-course calculus text contains the material presented in this chapter.

Answers To Odd-Numbered Exercises

Chapter 1

Section 1

1.1 a) $\{0\}$ b) $\{f, a, r, m\}$ c) $\{h, a, p, y\}$ d) $\{0\}$
e) $\{-2, -1, 0, 1, 2\}$ f) $\{-1, 0, 3\}$

1.3 a) $\{1, 3, 6\}$ b) $\{1, 2, 3, 4, 5, 6, 7, 8\}$ c) $\{2, 4, 5, 7, 8\}$
d) $\{2, 4, 5\}$ e) $\{7, 8\}$

1.5 a) $\{1, 2, 3, 4, 5\}$ b) $\{5, 6, 7, \ldots, 1000\}$ c) $\{1, 2, 3, 4\}$
d) $\{1001, 1002, \ldots, 2000\}$ e) $\{-5, -4, -3, -2, -1, 0, 1, \ldots, 2000\}$
f) $\{-5, -4, -3, -2, -1, 0, 6, 7, 8, \ldots, 1000\}$

1.7 a) $\varnothing, \{1\}, \{2\}, \{3\}, \{1, 2\}, \{1, 3\}, \{2, 3\}, \{1, 2, 3\}$. The improper subsets are $\varnothing$ and $\{1, 2, 3\}$.
b) $\varnothing, \{1\}, \{\{2, 3\}\}, \{1, \{2, 3\}\}$. The proper subsets are $\{1\}$ and $\{\{2, 3\}\}$.

1.9 a) Not well defined   b) Well defined
c) Well defined, it is $\varnothing$ d) Not well defined

1.11 a) Let $x \in A$. Now $A \subseteq B$, so $x \in B$, and then $B \subseteq C$, so $x \in C$. Hence $A \subseteq C$.
b) Let $A = \{a\}$, $B = \{\{a\}, b\}$, and $C = \{\{\{a\}, b\}, c\}$.

Section 2

2.1 "Q" stands for "quotient"; a rational number is a quotient of integers.

2.3 F T T F T F T T

2.5 $A \subseteq A$, $A \subset B$, $B \subseteq B$, $C \subseteq C$, $C \subset E$, $D \subseteq D$, $D \subset E$, $E \subseteq E$, $F \subset A$, $F \subset B$, $F \subset C$, $F \subset D$, $F \subset E$, $F \subseteq F$

2.7 a) $\{1, 2\}$ b) $\varnothing$ c) $\{-1, 0\}$ d) $\{1, 2\}$

2.9 (Sketches omitted)

Section 3

3.1 a) B b) $\{6, 7, 8, 9, 10\}$ c) U d) $\varnothing$
e) $\{2, 4, 6, 7, 8, 9, 10\}$ f) $\{7, 9\}$
g) $\overline{B \cup C} = \{6, 8, 10\}$, $\bar{B} \cup \bar{C} = \{2, 4, 6, 7, 8, 9, 10\}$
h) $\overline{A \cap C} = \bar{A} \cup \bar{C} = \{1, 3, 5, 6, 7, 8, 9, 10\}$
i) $\overline{B \cup C} = \bar{B} \cap \bar{C} = \{6, 8, 10\}$

3.3 a) $\bar{M} \cap Y$ b) $M \cap \bar{Y}$ c) $\bar{S}$ d) $\bar{S} \cup \bar{Y}$
e) $\bar{S} \cap M \cap Y$ f) $S \cap \bar{M} \cap \bar{Y}$

3.5 The regions shown numbered in Fig. 4.1 should be labeled as follows: 1, $A \cap B$; 2, $A \cap \bar{B}$; 3, $\bar{A} \cap B$; 4, $\bar{A} \cap \bar{B}$.

3.7 a) 5   b) 8   c) 0   d) 3

3.9 a) $A \subseteq B$ b) $B \subseteq A$

3.11 a) The region numbered 2 in Fig. 4.1 should be shaded.
b) $A - C = \{6, 8, 10\}$, $C - B = \{2, 4\}$, $A - A = \varnothing$
c) $(A \cup B) - C = \{6, 7, 8, 9, 10\}$, $A \cup (B - C) = \{2, 4, 6, 7, 8, 9, 10\}$. Yes.

3.13 a) The regions numbered 2 and 3 in Fig. 4.1 should be shaded.
b) $A + B = \{1, 2, 3, 4, 5, 6, 7, 8, 9, 10\}$, $B + C = \{2, 4, 7, 9\}$, $C + C = \varnothing$
c) $(A + B) \cup C = \{1, 2, 3, 4, 5, 6, 7, 8, 9, 10\}$,
$A + (B \cup C) = \{1, 3, 5, 6, 7, 8, 9, 10\}$. Yes.

3.15 a) $\mathbf{Z}^+$ b) $\{1\}$

3.17 a) $\{x \in \mathbf{R} \mid x > 0\}$ b) $\varnothing$

Section 4

4.1 a) 32   b) 64   c) 2^n

4.3 a) $\varnothing$ b) Regions 1, 2, 4, 5, 6, 7 of Fig. 4.2

4.5

4.7 Both sets comprise regions 1, 3, 4 of Fig. 4.2.

4.9 a) and b) Both sets comprise regions 1 and 2 of Fig. 4.1.

4.11 Both sets comprise regions 1, 2, 3, 4, 5, 6, 7 of Fig. 4.2.

4.13 $(\bar{A} \cap B) \cap (C \cup A) = (\bar{A} \cap B \cap C) \cup (\bar{A} \cap B \cap A)$
$$= (\bar{A} \cap B \cap C) \cup (\bar{A} \cap A \cap B)$$
$$= (\bar{A} \cap B \cap C) \cup (\varnothing \cap B) = (\bar{A} \cap B \cap C) \cup \varnothing$$
$$= \bar{A} \cap B \cap C = B \cap (C \cap \bar{A})$$

4.15 $\overline{A \cap (B \cup \bar{C})} = \bar{A} \cup \overline{(B \cup \bar{C})} = \bar{A} \cup (\bar{B} \cap \bar{\bar{C}}) = \bar{A} \cup (\bar{B} \cap C)$

4.17 $\overline{(B \cap \bar{C}) \cup (\bar{A} \cap \bar{C})} = \overline{(B \cap \bar{C})} \cap \overline{(\bar{A} \cap \bar{C})} = (\bar{B} \cup \bar{\bar{C}}) \cap (\bar{\bar{A}} \cup \bar{\bar{C}})$
$$= (\bar{B} \cup C) \cap (A \cup C) = (\bar{B} \cap A) \cup C$$

4.19 a) $\varnothing$
b) $(A \cap B \cap C) \cup (\bar{A} \cap B \cap C) \cup (A \cap B \cap \bar{C}) \cup (\bar{A} \cap \bar{B} \cap C)$
$$\cup (A \cap \bar{B} \cap \bar{C}) \cup (\bar{A} \cap B \cap \bar{C})$$

4.21 $A \cap (\bar{B} \cup C) = (A \cap \bar{B}) \cup (A \cap C)$

$\qquad = [(A \cap \bar{B}) \cap (C \cup \bar{C})] \cup [(A \cap C) \cap (B \cup \bar{B})]$

$\qquad = (A \cap \bar{B} \cap C) \cup (A \cap \bar{B} \cap \bar{C}) \cup (A \cap B \cap C)$

$\qquad\qquad\qquad\qquad\qquad\qquad\qquad\qquad \cup (A \cap \bar{B} \cap C)$

$\qquad = (A \cap B \cap C) \cup (A \cap \bar{B} \cap C) \cup (A \cap \bar{B} \cap \bar{C})$

Section 5

5.1 $n(A \cup B \cup C \cup D) = n(A) + n(B) + n(C) + n(D)$

$\qquad - n(A \cap B) - n(A \cap C) - n(A \cap D) - n(B \cap C)$

$\qquad - n(B \cap D) - n(C \cap D) + n(A \cap B \cap C)$

$\qquad + n(A \cap B \cap D) + n(A \cap C \cap D)$

$\qquad + n(B \cap C \cap D) - n(A \cap B \cap C \cap D)$

5.3 57

5.5 a) 37   b) 12   c) 15

5.7 a) 9   b) 17   c) 13

5.9 a) 14   b) 28   c) 0

5.11 a) You can't; the data is inconsistent.

b) Don't fire him! Send him to night school to learn about Venn diagrams.

5.13 It is impossible for $A \cap B \cap C$ to have more elements than $B \cap C$, since $(A \cap B \cap C) \subseteq (B \cap C)$.

Section 6

6.1 a) From each of three branches labeled 1, 2, 3, there emanate two branches labeled 1, 2.

b) From each of two branches labeled 1, 2, there emanate three branches labeled 1, 2, 3.

6.3 The stages of the trees are in opposite order.

6.5 a) 12   b) 12   c) 24   d) 64   e) 36

f) $(108)^3 = 1,259,712$

6.7 (Partial answers) a) A vertical line one unit to the right of the vertical number line

b) A vertical strip three units wide

c) A vertical line segment five units long

d) The "diagonal" line segment joining $(-1, -1)$ to $(1, 1)$

6.9 (Partial answers) a) A rectangle two units wide by one unit high

b) A horizontal line segment two units long

c) Three parallel vertical line segments two units long

d) Six points

6.11 Each point of $\mathbf{R}^3$ can be viewed as a point in Euclidean three-dimensional space. See Fig. 1.4 of the chapter on real analytic geometry.

6.13 a) $(1, -3)$ b) $(4, 2)$ c) $(\varnothing, 1)$ d) $(1, \{\varnothing\})$

6.15 $\{\{a\}, \{a, b\}, \{a, b, c\}\}$ **6.17** a) $\{(3, 2)\}$ b) $S_1 \cap S_2$

Section 7

7.1 a) 4   b) 6   c) 2   d) 8   e) 0   f) 5

7.3 a) A map   b) Not a map; two pairs have 1 as first member.
c) A map   d) Not a map; no pair has 1 as first member.

7.5 (Partial answers) a) $\{y \in \mathbf{R} \mid y \geq 0\}$ b) $\mathbf{R}$ c) $\mathbf{R}$
d) $\{y \in \mathbf{R} \mid y \geq 4\}$ e) $]0, 1]$

7.7 r

7.9 a) 3   b) -3 c) -3

7.11 a) Map; Domain $= \{(1, 2), 1\}$, Range $= \{3, (2, 3)\}$
b) Not a map; second element is not an ordered 2-tuple.
c) Map; Domain $= \{(1, 2), 3, (2, 4, 1)\}$, Range $= \{(1, 2), (1, 2, 4), 3\}$
d) Not a map; two 2-tuples have 1 as first member.
e) Map; Domain $= \{(1, 2), (2, 1), 2\}$, Range $= \{3, (1, 3)\}$
f) Not a map; two 2-tuples have $(1, 2)$ as first member.

Section 8

8.1 $5! = 120$ **8.3** (Partial answer) $3! = 6$ **8.5** (Partial answer) 5

8.7 61   **8.9** 24   **8.11** $4, 8, 16, 2^n$

8.13 15   **8.15** 6

8.17 a) 3^{31} b) $3(2^{30})$ c) 6

8.19 $12 \cdot 11 \cdot 10 = 1320$

Section 9

9.1 a) $2^7 = 128$ b) 126

9.3 a) $2^3 = 8$ b) $2^6 - 2^3 = 56$ c) $2^6 - 2^3 - 2^3 + 1 = 49$

9.5 a) $\{\{(a, 1), (b, 1), (c, 1)\}\}$
b) $\{\{(1, a)\}, \{(1, b)\}, \{(1, c)\}\}$

9.7 a) $2^2 = 4$ b) $6^3 = 216$ c) $2(2^3) = 16$ d) $2^6 = 64$

9.9 $\{(1, 1), (2, 1), (3, 0), (4, 0), (5, 1)\}$

9.11 $\{(1, 1), (2, 1), (3, 1), (4, 1), (5, 1)\}$

9.13 $S = \{1, -1\}$, $A = \{3, 1, 2, 5, -1\}$

9.15 $\varnothing \times A = \varnothing$, and the subset $\varnothing$ of $\varnothing \times A$ has the property required. Thus $A^\varnothing = \{\varnothing\}$, so $n(A^\varnothing) = n(\{\varnothing\}) = 1 = (n(A))^0 = n(A)^{n(\varnothing)}$.

Section 10

10.1 $\binom{12}{4} = 495$ **10.3** $\binom{6}{4} + \binom{6}{3}\binom{6}{1} = 135$

10.5 a) $\binom{7}{3}\binom{5}{2} = 350$ b) 270   **10.7** $3^8 = 6,561$

10.9 There are r choices for the box in which to put the first of the n elements, *and* then r choices for the box for the second, etc., giving r^n choices in all, by Theorem 8.2.

10.11 By Theorem 9.1, a set of n elements has 2^n subsets. However, each subset either has zero elements *or* one element *or* two elements *or* . . . *or* $n - 1$ elements *or* n elements, so by Theorem 10.1, the total number of subsets is also the sum

$$\binom{n}{0} + \binom{n}{1} + \binom{n}{2} + \ldots + \binom{n}{n-1} + \binom{n}{n}.$$

10.13 a) $\binom{7}{3}\binom{4}{2} = 210$ b) $\binom{5}{3}\binom{4}{2} = 60$

10.15 a) $6! = 720$ b) $\binom{6}{2} = 15$ c) $\binom{6}{3}\binom{3}{2} = 60$

10.17 Let U have one red element and $n - 1$ black elements. Then by Exercise 10.16, U has $2^{n-1}/2$ subsets containing an even number of black elements but not the red element. Also by Exercise 10.16, U has $2^{n-1}/2$ subsets containing an odd number of black elements and the red element. Thus U has

$$\frac{2^{n-1}}{2} + \frac{2^{n-1}}{2} = 2^{n-1}$$

subsets with an even number of elements.

Chapter 2

Section 1

1.1 a) $\frac{2}{3}$ b) 0   c) 1   d) $\frac{1}{3}$ e) $\frac{1}{3}$ f) $\frac{2}{3}$ g) $\frac{2}{3}$

1.3 No

1.5 a) No; the case of one head and one tail is not covered.
b) Yes
c) No; the case of one head and one tail is not covered.
d) Yes
e) No; the case of one head and one tail is covered twice.

1.7 a) Yes   b) Yes
c) No; an even number might be either ≤ 4 or ≥ 4.   d) Yes
e) No; for the second and third outcomes, e might or might not hold.

1.9 $\text{pr}[1] = \text{pr}[3] = \text{pr}[5] = \frac{1}{9}$; $\text{pr}[2] = \text{pr}[4] = \text{pr}[6] = \frac{2}{9}$

1.11 a) $\text{pr}[R] = \frac{1}{2}$, $\text{pr}[G] = \frac{1}{3}$, $\text{pr}[Y] = \frac{1}{6}$
b) All outcomes have weight $\frac{1}{6}$.
c) $\text{pr}[R] = \frac{1}{2}$; all other outcomes have weight $\frac{1}{6}$.
d) $\text{pr}[R_1] = \frac{1}{6}$, $\text{pr}[G] = \frac{1}{3}$, $\text{pr}[other] = \frac{1}{2}$

1.13 $\frac{1}{2}$

1.15 $\frac{3}{5}$

1.17 a) Many people guess about 180.
b) Assuming that birthdays are as likely to occur on one calendar day as another, when $n = 22$, $\text{pr}[e] = 0.476$, and when $n = 23$, $\text{pr}[e] = 0.507$.

Section 2

2.1 $\frac{3}{8}$ **2.3** $\frac{1}{6}$ **2.5** $\frac{5}{36}$

2.7 a) $\frac{1}{2}$ b) $\frac{1}{4}$ c) $\frac{1}{2}$ d) $\frac{5}{16}$

2.9 The same number of H as T can happen in $\binom{2n}{n}$ of the 2^{2n} ways, so

$$\text{pr}[\textit{same number of } H \textit{ as } T] = \frac{\binom{2n}{n}}{2^{2n}} \; .$$

By symmetry, for a fair coin, pr[*more H than T*] = pr[*more T than H*], so each has half the remaining weight.

2.11 $\dfrac{\binom{13}{5}\binom{13}{3}\binom{26}{5}}{\binom{52}{13}}$ **2.13** $\dfrac{13 \cdot 48}{\binom{52}{5}}$

2.15 With reference to Fig. 4.1 of Chapter 1, with A replaced by E_1 and B by E_2, we see that $m(E_1 \cup E_2)$ is the total weight of regions 1, 2, and 3. This is the total weight of regions 1 and 2 plus the total weight of regions 1 and 3, minus the weight of region 1.

Section 3

3.1 $\frac{2}{3}$ **3.3** $\frac{1}{2}$ **3.5** $\frac{4}{33}$

3.7 $\dfrac{\binom{13}{4}\binom{13}{3}}{\binom{26}{7}}$

3.11 a) Independent   b) Not independent

Section 4

4.1 $\frac{1}{16}$ **4.3** $\frac{10}{13}$ **4.5** $\frac{1}{2}$ **4.7** $\frac{1}{2}$

4.9 $\frac{144}{343}$

4.11 a) $\frac{1}{2}$ b) $\frac{1}{2}$ c) 0

Section 5

5.1 $\frac{15}{64}$ **5.3** 3   **5.5** $(\frac{5}{6})^4 = \frac{625}{1296}$

5.7 $\frac{99}{128}$ **5.9** $\frac{1}{2}$ **5.11** 0

5.13 $\frac{36}{125}$

Section 6

6.1 90 cents   **6.3** $\frac{7}{3}$ **6.5** $\frac{5}{2}$

6.7 a) $\frac{11}{3}$ b) $\frac{55}{3}$

6.9 a) 1   b) $\frac{1}{2}$ c) 1   d) It is the sum of the answers.

6.11 $\frac{40}{3}$ cents

Section 7

7.1 $\frac{1}{2}$ **7.3** $\frac{65}{27}$ **7.5** $\frac{1}{3}$ **7.7** $(\frac{37}{48})^2$

7.9 By Theorem 7.2, the particle will return to its initial position after some finite number of steps with probability 1. Think of a new walk starting each time the particle returns to its initial position. By Theorem 7.2, the particle will always return to start a new walk with probability 1. If the process continues indefinitely, the particle must return an infinite number of times with probability 1.

7.11 The particles are five units apart at the start. It is easy to see that at any step the number of units distance between the particles either remains the same, increases by 2, or decreases by 2. In particular, the distance between them is always odd, so it can never be zero.

7.13 a) $A + B$ dollars
b) $X + 1$ with probability p, and $X - 1$ with probability $1 - p$
c) This is obvious.

Section 8

8.1 a) 0.118   b) 0.236   c) 0.532   d) 0.150   e) 0.976
f) 0.045   g) 0.018   h) 0.919

8.3 0.818   **8.5** 0.012   **8.7** 0.816   **8.9** 0.023

8.11 0.0013

Section 9

9.1 [0.58, 0.64]   **9.3** 722,500

9.5 At least $\frac{495}{1000}$ and at most $\frac{545}{1000}$ **9.7** No   **9.9** 184

Chapter 3

Section 1

1.1 a) Euclidean 4-space is $\mathbf{R}^4$.
b) The distance between (a_1, a_2, a_3, a_4) and (b_1, b_2, b_3, b_4) is

$$\sqrt{(a_1 - b_1)^2 + (a_2 - b_2)^2 + (a_3 - b_3)^2 + (a_4 - b_4)^2}.$$

c) The 3-sphere in $\mathbf{R}^4$ with center (a_1, a_2, a_3, a_4) and radius $r > 0$ is

$\{(x_1, x_2, x_3, x_4) \in \mathbf{R}^4 \mid (x_1 - a_1)^2 + (x_2 - a_2)^2 + (x_3 - a_3)^2 + (x_4 - a_4)^2 = r^2\}.$

d) The 4-ball in $\mathbf{R}^4$ with center (a_1, a_2, a_3, a_4) and radius $r > 0$ is

$\{(x_1, x_2, x_3, x_4) \in \mathbf{R}^4 \mid (x_1 - a_1)^2 + (x_2 - a_2)^2 + (x_3 - a_3)^2 + (x_4 - a_4)^2 \le r^2\}.$

1.3 a) 3   b) 5   c) $\sqrt{2} + \pi$

1.7 (Partial answer) a) The plane through $(2, 0, 0)$ parallel to the x_2- and x_3-coordinate plane
b) The plane through $(0, 0, 3)$ parallel to the x_1- and x_2-coordinate plane
c) The plane containing the x_3-axis and passing through the point $(1, 1, 0)$

1.9 a) $(2, 1, -4)$ b) $(1, -\pi, -\sqrt{2})$ c) $(-1, 0, -3)$
d) $(-1, 2, 2)$

1.11 (Partial answer) a) Two parallel planes, one through $(0, 1, 0)$ and one through $(0, -1, 0)$, and each parallel to the x_1- and x_3-coordinate plane
b) The infinite right circular cylinder with radius 1 and x_3-axis down its center
c) The 2-sphere with center at the origin and radius 1
d) The line joining $(0, 0, 0)$ and $(1, 1, 1)$

1.13 $\{(x_1, x_2, x_3) \in \mathbf{R}^3 \mid (x_1 - 1)^2 + x_2{}^2 + (x_3 + 3)^2 \le 4\}$

1.15 $(x_1 - 0)^2 + (x_2 + 1)^2 + (x_3 - 3)^2 = 16$. Center $(0, -1, 3)$. Radius 4

Section 2

2.1 a) $(-3, 0)$ to $(-3, 1)$ to $(3, 1)$ to $(3, 0)$; 8 units
b) $(-3, 0)$ to $(-3, 0.05)$ to $(3, 0.05)$ to $(3, 0)$

2.3 a) $(0, 0, 0, 0)$ to $(0, 0, 0, 1)$ to $(1, -1, 1, 1)$ to $(1, -1, 1, 0)$; $2 + \sqrt{3}$

b) $(0, 0, 0, 0)$ to $\left(0, 0, 0, \dfrac{2 - \sqrt{3}}{2}\right)$ to $\left(1, -1, 1, \dfrac{2 - \sqrt{3}}{2}\right)$ to $(1, -1, 1, 0)$

2.5 Turn the left glove over via $\mathbf{R}^3$, and it becomes the right glove.

2.7 He could escape from any two-dimensional box, manufacture a pair of gloves from two left gloves, etc.

2.9 $(0, 0, 0, 0, 0, 0, 0, 0, 0, 0, 0)$ to $(0, 0, 0, 0, 0, 0, 0, 0, 0, 0, 1)$ to
$(1, -1, 1, -1, 1, -1, 1, -1, 1, -1, 1)$ to $(1, -1, 1, -1, 1, -1, 1, -1, 1, -1, 0)$;
$2 + \sqrt{10}$

2.11 a) If a 3-sphere in $\mathbf{R}^4$ is intersected with
  i) a nontangent line, a 0-sphere results,
  ii) a nontangent plane, a 1-sphere results,
  iii) a nontangent 3-sphere, a 2-sphere results.
b) A 2-sphere
c) Such an intersection gives an $(s - 1)$-sphere.

2.13 He sees one or more objects which may approach or recede, change shape, color etc., as time passes. (Note that the change in appearance is really due to a change in position of the three-dimensional bug, rather than to the passing of time. The reason the two-dimensional bug may see several objects rather than one is due to the fact that only part of the three-dimensional bug is in the two-dimensional bug's world at a certain moment; for example, perhaps only parts of the three-dimensional bug's feet are in the two-dimensional bug's world at the start.)

Section 3

3.1 (Sketch omitted)

3.3 a) $\sqrt{17}$ b) $\sqrt{17}$ c) $2\sqrt{17}$ d) 15

3.5 a) 3   b) $-\frac{6}{5}$ c) 0   d) Impossible

3.7 $\dfrac{1}{\sqrt{\pi}}, 0, \dfrac{-1}{\sqrt{\pi}}, \dfrac{3}{\sqrt{11}}$

3.9 $\dfrac{1}{\sqrt{2}}, 0, \dfrac{1}{\sqrt{2}}$, $\dfrac{1}{\sqrt{5}}, \dfrac{2}{\sqrt{5}}, 0$

3.11 $<3, -2, 4> - <1, -1, 4> = <2, -1, 0>$
$$= <-2, 1, 6> = <-4, 2, 6>.$$

Also

$$<3, -2, 4> - <-2, 1, 6> = <5, -3, -2>$$
$$= <1, -1, 4> - <-4, 2, 6>.$$

Section 4

4.1 (Partial answer) $x_1 = 3 - 8t, \, x_2 = -2 + 4t$

4.3 $x_1 = 2 - t, \, x_2 = -1, \, x_3 = 4t, \, x_4 = 3 - 3t$

4.5 $x_1 = 5 + t, \, x_2 = -1 + 2t$

4.7 a) Intersect at $(-1, 4)$; not orthogonal   b) Parallel
c) Intersect at $(0, -2)$; orthogonal

4.9 The lines are the same; $\{(5 - 3t, -1 + t) \mid t \in \mathbf{R}\}$.

4.11 a) $(\tfrac{1}{2}, \tfrac{3}{2})$ b) $(\tfrac{3}{2}, -2, \tfrac{5}{2})$ c) $(2, \tfrac{1}{2}, -\tfrac{1}{2}, \tfrac{3}{2})$

4.13 $(-\tfrac{3}{2}, -\tfrac{1}{2}, \tfrac{11}{4}, 0)$

4.15 The line segment consists of all $(x_1, x_2, \ldots, x_n)$ such that $x_i = u_i + t(b_i - a_i)$ for $0 \le t \le 1$. But $a_i + t(b_i - a_i) = (1 - t)a_i + b_i$.

Section 5

5.1 a) $\mathbf{R}$ b) A point of $\mathbf{R}$

5.3 $3x_1 - 2x_2 + 7x_3 = 39$ **5.5** $3x_1 - 7x_2 + 3x_3 = 0$

5.7 $x_1 = -2 + t, \, x_2 = 1 - 2t, \, x_3 = t, \, x_4 = 5 + 4t$

5.9 $(2, 3, 4, -2, 4)$

5.11 The hyperplane is a line parallel to the x_1-axis and orthogonal to the x_2-axis.
b) The hyperplane is a plane parallel to the x_1-axis and orthogonal to the x_2- and x_3-coordinate plane.
c) The hyperplane is a 3-space in $\mathbf{R}^4$ parallel to the x_1-axis and orthogonal to the x_2-, x_3- and x_4-coordinate 3-space.
d) The hyperplane is a plane parallel to the x_1- and x_2-coordinate plane and orthogonal to the x_3-axis.
e) The hyperplane is a 3-space in $\mathbf{R}^4$ parallel to the x_1- and x_2-coordinate plane and orthogonal to the x_3- and x_4-coordinate plane.

5.13 $x_1 + x_2 + x_3 + x_4 = 1$

Section 6

6.1 $x_1 + 3x_2 = 18, \, 2x_2 - x_3 = 9$ (also $2x_1 + 3x_3 = 9$)

6.3 $x_1 + 3x_2 = 11$ **6.5** $x_1 = -2 + t, \, x_2 = 4 - 2t$

6.7 $y = 10x - 13$ **6.9** $x_1 = t, \, x_2 = -4 + 2t$

6.11 $y = -\tfrac{3}{2}x + \tfrac{5}{2}$

Chapter 4

Section 1

1.1 $V = x^3$

1.3 $A = s^2/4\pi$

1.5 a) 0   b) 0   c) 1   d) Undefined   e) Undefined
f) 3   g) -2 h) Undefined

1.7 a) Domain: $\{x \in \mathbf{R} \mid x \geq -1\}$; Range: $\{x \in \mathbf{R} \mid x \geq 0\}$
b) Domain: $\mathbf{R}$; Range: $\{x \in \mathbf{R} \mid x \geq 1\}$
c) Domain: $\{x \in \mathbf{R} \mid x \leq -1$ or $x \geq 1\}$; Range: $\{x \in \mathbf{R} \mid x \geq 0\}$
d) Domain: $\{x \in \mathbf{R} \mid x \geq -\frac{3}{2}\}$; Range: $\{x \in \mathbf{R} \mid x \geq 0\}$

1.9 (Sketches omitted)

1.11 a) $\mathbf{R}^4$ b) $\{(x_1, x_2, x_3, x_4) \in \mathbf{R}^4 \mid x_4 \geq 0\}$ c) $\mathbf{R}^4$
d) On or outside the 3-sphere with center at the origin and radius 4.

Section 2

2.1 To define the sum $f + g$, $f(x) + g(x)$ has to make sense for x in the domain of f and g, i.e., you have to have a concept of adding an element of the range of f to an element of the range of g. Similar considerations hold for $f - g$, fg, and f/g.

2.3 a) $\mathbf{R}$ b) $\{x \in \mathbf{R} \mid x \neq -\frac{1}{2}\}$ c) $\mathbf{R}$ d) $\{x \in \mathbf{R} \mid x \neq -1, 0, 1\}$

2.5 a) $\mathbf{R}^2$ b) $\{(x, y, z) \in \mathbf{R}^3 \mid z \neq 0\}$ c) $\mathbf{R}^3$
d) $\{(x, y, z) \in \mathbf{R}^3 \mid xyz \neq 0\}$ e) $\{(x, y, z) \in \mathbf{R}^3 \mid z \neq 1\}$

2.7 (Sketch omitted)

Section 3

3.1 a) Domain: $\{2, 1, 0\}$; Range: $\{-3, 1\}$
b) Domain: $\{1, 2, 3\}$; Range: $\{1\}$
c) Domain: $\{(2, 0), (1, 1), (2, 1)\}$; Range: $\{1, -2, 4\}$
d) Domain: $\{0, 1, 2\}$; Range $\{(0, 1), (1, -2), (1, 4)\}$

3.3 a) 1   b) $\frac{1}{2}$ c) $-\frac{3}{2}$

3.5 (Sketches omitted)

3.7 (Partial answers)   a) A plane
b) A bowl with bottom point $(0, 0, -1)$, having increasing circular cross sections as z increases, and extending upward without end.
c) A plane containing the y-axis.

3.9 a) $\{\begin{smallmatrix}1\\3\end{smallmatrix}\}$ b) $\{1, -1\}$ c) $\{-1\}$

3.11 (Partial answer) The level sets are concentric circles with the origin as center.

Section 4

4.1 a) -2 b) -8 c) 4
4.3 a) 7   b) 18

4.5 a) $(-1, 2, 1)$ b) $(6, -8, 0)$ c) $(16, -20, 2)$

4.7 $f(-1 \prec 1 \succ) = f(\prec -1 \succ) = \sqrt{(1-)^2} = 1$, while $-1 f(\prec 1 \succ) = -1 \cdot 1 = -1$.

4.9 Yes, for $f(\prec a, b \succ + \prec c, d \succ) = 0 = 0 + 0 = f(\prec a, b \succ) + f(\prec c, d \succ)$ and $af(\prec b, c \succ) = a \cdot 0 = 0 = f(\prec ab, ac \succ)$ for all $a, b, c, d \in \mathbf{R}$.

4.11 Yes, for

$$f(\prec a, b, c \succ + \prec d, e, f \succ) = \prec 0, 0, 0, 0 \succ = \prec 0, 0, 0, 0 \succ + \prec 0, 0, 0, 0 \succ$$
$$= f(\prec a, b, c \succ) + f(\prec d, e, f \succ),$$

and

$$af(\prec b, c, d \succ) = a \prec 0, 0, 0, 0 \succ = \prec 0, 0, 0, 0 \succ = f(\prec ab, ac, ad \succ)$$

for all $a, b, c, d, e, f \in \mathbf{R}$.

4.13 Now $g(\prec x_1, \ldots, x_n \succ + \prec y_1, \ldots, y_n \succ) = g(\prec x_1 + y_1, \ldots, x_n + y_n \succ) = b_1(x_1 + y_1) + \cdots + b_n(x_n + y_n) = (b_1 x_1 + \cdots + b_n x_n) + (b_1 y_1 + \cdots + b_n y_n) = g(x_1, \ldots, x_n) + g(y_1, \ldots, y_n)$. Also for $a \in \mathbf{R}$, we have

$$g(a \prec x_1, \ldots, x_n \succ) = g(\prec ax_1, \ldots, ax_n \succ)$$
$$= b_1 ax_1 + \cdots + b_n ax_n$$
$$= a(b_1 x_1 + \cdots + b_n x_n) = ag(x_1, \ldots, x_n).$$

4.15 a) $(-1, 2, 1)$ b) $(-1, 3, 3)$ c) $(-2, -1, 4)$

4.17 $(4, 0, 0)$, $3 \, dx + 4 \, dy + dz = 0$

Section 5

5.1 (Sketches omitted)

5.3 a) $\frac{4}{5}$ b) 0   c) Does not exist   d) 0   e) -1

5.5 Yes, for $\lim_{x \to 3} f(x) = 6 = f(3)$, and f is continuous as a rational function at all points other than 3 in $\mathbf{R}$.

5.7 Let $\epsilon > 0$ be given and let $\delta_\epsilon = \epsilon/5$. If $-3 - \delta_\epsilon < x < -3 + \delta_\epsilon$, then $-15 - 5\delta_\epsilon < 5x < -15 + 5\delta_\epsilon$, so $-14 - 5\delta_\epsilon < 5x + 1 < -14 + 5\delta_\epsilon$. But $5\delta_\epsilon = \epsilon$, and $-14 = f(-3)$, so $f(-3) - \epsilon < f(x) < f(-3) + \epsilon$, provided $-3 - \delta_\epsilon < x < -3 + \delta_\epsilon$.

5.9 a) For at least one $\epsilon > 0$, there does not exist $\delta_\epsilon > 0$.

b) For at least one apple blossom, there is no apple.

d) Find an $\epsilon > 0$ such that for all $\delta_\epsilon > 0$, there is some $x \neq a$ inside the sphere with center a and radius δ_ϵ such that $f(x)$ is outside the sphere with center c and radius ϵ.

Chapter 5

Section 1

1.1 a) $\dfrac{(1 + dx)^2 - 1}{dx}$ b) $\dfrac{(dx)^2}{dx}$ c) $\dfrac{\dfrac{1}{2 + dx} - \dfrac{1}{2}}{dx}$

d) $\dfrac{\sqrt{4 + dx} - 2}{dx}$ e) $\dfrac{\sqrt{9 + 2dx} - 3}{dx}$ f) $\dfrac{(x_0 + dx)^2 - x_0{}^2}{dx}$

1.3 a) $\dfrac{(-1 + dx)^2 - 1}{dx}$ b) -2 c) $f(-1.05) \approx 1.1$

d) $(-1.05)^2 = 1.1025$; the error is 0.0025.

1.5 a) $\dfrac{\sqrt{4 + dx} - \sqrt{4}}{dx}$ b) $\frac{1}{4}$ c) 1.995

1.7 $\dfrac{-1}{x_0{}^2}$

1.9 a) The graph of $4x + 3$ is a straight line, so a linear approximation is exact.
b) 4
1.11 $y - y_0 = f'(x_0)(x - x_0)$, or $y - f'(x_0)x = y_0 - f'(x_0)x_0$
1.13 $y = -x + 2$ **1.15** $x = 1 + t, y = 5 + 2t$

Section 2

2.1 a) At time $t = 10$ sec, Bill's speed is 15 ft/sec.
b) When exactly one minute has elapsed, Mike's speed is 12 ft/sec.
c) When exactly one-half minute has elapsed, Bill is running half again as fast as Mike.

2.3 20 mph

2.5 dA/dx is the instantaneous rate of change of the area of the triangle per unit change in the length x of a side.

2.7 a) A horizontal line
b) A straight line with direction vector $<1, v_0>$, that is, with slope v_0
2.9 a) 48 in³/sec   b) 300 in³/sec

Section 3

3.1 a) $7x^6$ b) $30x^5$ c) $12x^3 + 14x$ d) $-\dfrac{3}{x^2}$

e) $-\dfrac{10}{x^3} - \dfrac{1}{x^2}$ f) $6x^2 - \dfrac{9}{x^4}$ g) $16x + 3 + \dfrac{7}{x^2}$ h) $3 - \dfrac{14}{x^2} + \dfrac{21}{x^4}$

3.3 a) $30(3x + 1)^9$ b) $5(x^2 - 5x^3)^4 \cdot (2x - 15x^2)$ c) $\dfrac{5}{2\sqrt{5x}}$

d) $\dfrac{3x}{\sqrt{3x^2 + 1}}$ e) $-(4x^3 + 9x)^{-4/3} \cdot (4x^2 + 3)$

f) $5[(x^2 + 4x)^3 - 9x^2]^4 \cdot [3(x^2 + 4x)^2 \cdot (2x + 4) - 18x]$

3.5 70.49 **3.7** 8.69

3.9 $\dfrac{3}{64\pi}$ ft/min

3.11 a) $\dfrac{1}{(x+1)^2}$ **b)** $\dfrac{-4x}{(x^2-2)^2}$ **c)** $(x-2)^2 \cdot (2x+3)(10x+1)$

d) $\dfrac{3x^2-6x+6}{(x^2-2)^2}$ **e)** $\dfrac{(x+2)^2 \cdot (x-7)}{(x-1)^3}$

Section 4

4.1 -8 **4.3 a)** -4 **b)** $a=3, b=-12$ **c)** No

4.5 Let P be the perimeter, x the width, and y the length. We wish to maximize $A = xy$. Now $P = 2x + 2y$, so $A = x(\frac{1}{2}(P-2x)) = \frac{1}{2}(Px - 2x^2)$. Then $dA/dx = \frac{1}{2}(P-4x)$, so $dA/dx = 0$ when $x = P/4$. Therefore $y = P/4$ also. Since dA/dx changes from positive to negative as x increases at $P/4$, these values for x and y give a maximum area.

4.7 $x=2, y=4$

4.9 $\dfrac{r}{\sqrt{2}}$ by $2\dfrac{r}{\sqrt{2}}$

Section 5

5.1 a) $0,0,0,$ **b)** $20x^3 - 6, 60x^2, 120x$ **c)** $\dfrac{2}{x^3}, \dfrac{-6}{x^4}, \dfrac{24}{x^5}$

d) $6x - \dfrac{18}{x^4}, 6 + \dfrac{72}{x^5}, -\dfrac{360}{x^6}$

5.3 $x^5 + \dfrac{x}{4} - 30$ **5.5** $x = 2t^4 - 3t^2 + 4$ **5.7** 196 ft

5.9 $v = -32t + v_0, x = -16t^2 + v_0 t + s_0$

5.11 Let $f(x) = \begin{cases} 1 & \text{for } x > 0, \\ -1 & \text{for } x < 0. \end{cases}$

Section 6

6.1 3, 16

6.3 a) y, x **b)** $2x, 6y$ **c)** $y^3 + 2xy^2, 3xy^2 + 2x^2y$

d) $\dfrac{1}{y}, -\dfrac{x}{y^2}$ **e)** $\dfrac{2x+3}{y^2}, -\dfrac{x^2+3x+1}{y^2}$ **f)** $y^2 + \dfrac{6x}{y^3}, 2xy - \dfrac{9x^2}{y^4}$

6.5 Both third partial derivatives are $12xy - 6/y^4$.

6.7 $x_1 + x_2 - x_3 + x_4 = 4$. [Don't use calculus! An (orthogonal) direction vector for the hyperplane is directed along the radius, so $\langle 1, 1, -1, 1 \rangle$ is a direction vector for the hyperplane.]

6.9 $x = 1 + 8t, y = 2 + 27t, z = 23 - t$

Section 7

7.1 1.975   **7.3** 7

7.5 a) $x = 1, y = 2, z = \frac{5}{4}$ b) $\frac{15}{4}$

c) $z = t^4 + \dfrac{1}{(t + 1)^2}$ d) $\dfrac{15}{4}$

7.7 103 in³

7.9 The force is decreasing at the rate of $20G$ units per unit time.

Chapter 6

Section 1

1.1 $S_2 = 5, s_2 = 1$

1.3 a) (Sketch omitted)   b) -1 c) 1   d) 0

1.5 (Approximate answers) $S_4 = 0.7595$, $s_4 = 0.6345$

1.7 20

1.9 The first trapezoid has area

$$\frac{b - a}{n}\left(\frac{y_0 + y_1}{2}\right),$$

the next has area

$$\frac{b - a}{n}\left(\frac{y_1 + y_2}{2}\right),$$

etc. Adding these n quantities and factoring out $(b - a)/2n$, we obtain the given formula.

1.11 (Sketch omitted)

1.13 Let M_i be the maximum value of f in the ith interval for S_n. This interval is divided into two equal pieces when the sum S_{2n} is computed, and the maximum value of f in each piece is at most M_i. Thus these two intervals contribute no more to S_{2n} than the whole ith interval contributes to S_n.

Section 2

2.1 a) $2x + c$ b) c c) $\frac{1}{2}x^2 + c$ d) $x^3 + x + c$
e) $\frac{1}{5}x^5 + x^3 - \frac{5}{2}x^2 + 7x + c$ f) $\frac{2}{3}x^{3/2} + c$

2.3 a) $\frac{5}{4}$ b) $\frac{9}{2}$ c) 0   d) $\frac{1}{5}$ e) $\frac{8}{3}$

2.5 (Sketch omitted) $\frac{8}{5}$

2.7 For $n \neq -1$, we have

$$\int_{-a}^{a} x^n = \left.\frac{x^{n+1}}{n + 1}\right]_{-a}^{a} = \frac{1}{n + 1}[a^{n+1} - (-a)^{n+1}].$$

If n is odd so that $n + 1$ is even, then $(-a)^{n+1} = a^{n+1}$, so $\int_{-a}^{a} x^n = 0$.

2.9 a) The area from a to c under f is the area from a to b plus the area from b to c.
b) The graph of $-f$ is symmetric with the graph of f, but it is on the other side of the x-axis.
c) The height to the graph of $f + g$ is the sum of the heights to the graphs of f and g.
d) The height to the graph of cf is c times the height to the graph of f.

2.11 a) Let $F' = f$. Then $F(b) - F(a) = -(F(a) - F(b)) = -\int_b^a f$.
b) If $\int_a^b f + \int_b^c f = \int_a^c f$ for all a, b, $c \in \mathbf{R}$, then $\int_a^b f + \int_b^a f = \int_a^a f = 0$, so $\int_a^b f = -\int_b^a f$.

Section 4

4.1 $\frac{32}{3}$ **4.3** $\frac{1}{6}$

4.5 $\frac{4}{15}$ **4.7** $\frac{44}{15}$

4.9 $\frac{9}{2}$

Section 5

5.1 $\frac{16}{15}\pi$ **5.3** $\frac{16}{3}\pi$

5.5 $V = \int_0^h \pi \left(\frac{r}{h}x\right)^2 dx = \pi \left. \frac{r^2}{h^2}\frac{x^3}{3}\right]_0^h = \pi \frac{r^2}{h^2}\frac{h^3}{3} = \frac{\pi}{3}r^2 h$

5.7 $\dfrac{\pi}{6}$

Section 6

6.1 64 ft lbs   **6.3** $\frac{3}{4}$

6.5 $\frac{6}{7}$ **6.7** $12{,}480\pi$

Index